Dark Matter

Dark Matter
The New Science of the Microbiome

DR JAMES KINROSS

PENGUIN LIFE

AN IMPRINT OF

PENGUIN BOOKS

PENGUIN LIFE

UK | USA | Canada | Ireland | Australia
India | New Zealand | South Africa

Penguin Life is part of the Penguin Random House group of companies whose addresses can be found at global.penguinrandomhouse.com.

First published 2023
001

Copyright © Dr James Kinross, 2023

The moral right of the author has been asserted

Set in 12/14.75pt Dante MT Std
Typeset by Jouve (UK), Milton Keynes
Printed and bound in Great Britain by Clays Ltd, Elcograf S.p.A.

The authorized representative in the EEA is Penguin Random House Ireland, Morrison Chambers, 32 Nassau Street, Dublin D02 YH68

A CIP catalogue record for this book is available from the British Library

Hardback ISBN: 978–0–241–54397–9
Trade paperback ISBN: 978–0–241–54398–6

www.greenpenguin.co.uk

Penguin Random House is committed to a sustainable future for our business, our readers and our planet. This book is made from Forest Stewardship Council® certified paper.

For Sophie, Jacomo and Liberty

Contents

Prologue: Scrubbing Up   ix
Introduction: Good Sh*t   xv

PART ONE – THE MICROBIOME

1 A Library of the Known Microbial Universe   3
2 Midlife Crisis   19
3 How Microbes Kill Us   39
4 Inflammation   54
5 Sex and Bugs and Oestradiol   71
6 The Big Bang   91
7 Symbiotic Sentience   107
8 How to Live to One Hundred   124

PART TWO – THE EXPOSOME

9 The Global Microbiome   145
10 The War on Bugs   159
11 A Biotic Life   174
12 The Drugs Don't Work   191
13 The Microbiome Café   206

PART THREE – THE METAHUMAN

14 Extreme Phenotypes   235
15 Hacking the Symbiont   255
16 The Trillion-Dollar Product   269

Epilogue: How to Save the Microbiome   275
Notes   279
Glossary   309
Acknowledgements   313
Index   315

Prologue: Scrubbing Up

12.46 p.m. GMT on 11 September 2001: I had just finished my first ever appendicectomy in operating room number three at St Mary's Hospital, London, and I was ecstatic. After six long years at medical school, I was finally doing something of actual use. I didn't know that over the next twenty years I would spend many long nights guddling around the dark recesses of people's abdomens trying to find this elusive source of pain. I didn't know that I would become obsessed by the bacteria that live in the gut and their critical importance to the workings of our fragile bodies. And I didn't know that there comes a point in every surgeon's career when they conclude that it would be better for all concerned if there was no need for surgery at all.

You may have lost your appendix to a surgeon like me, and you may have been told that the appendix is an organ without a function – an evolutionary footnote. It's true that you can live perfectly happily without an appendix, but it is most certainly not useless. Every one of our organs, no matter how ridiculous it might look or sound or feel, has evolved and stayed with us for a reason. It's not mere coincidence or bad luck that people who have their appendix taken out as children have an increased risk of developing inflammatory bowel disease, *Clostridium difficile* infections, colorectal cancer and even irritable bowel syndrome (IBS).[1] Just because we haven't been able to define its purpose, we foolishly assume the appendix has none. But the humble appendix contains a multitude of microscopic life forms that were, until recently, completely alien to us. Their discovery means that this enduring mystery of modern science might not remain a mystery for much longer.

On that Tuesday in 2001 I walked out of the operating room and into the coffee area on the fourth floor of the hospital at 1.03 p.m., just as the second plane hit the South Tower of the World Trade

Center in New York. The whole world stopped, and we silently watched as 2,996 people were murdered. The hospital is on the flight path to Heathrow Airport. Later that night I stood staring out of the ward's window, marvelling at the empty sky, and felt the sense of dread that hung over the city. No one could have imagined how bloody the vengeance for this atrocity would be. The war on terror would cost 900,000 lives and US $8 trillion, displacing thirty-eight million people and their microbes across the globe.[2] The supersized, ready-to-eat, buy-one-get-one-free, instantly delivered digital culture that rode the bow-wave of the global military offensive homogenized the diets and lives of people who did not want it. It was a harmful acceleration of a globalized way of life that had its roots in the Second World War and that has left microbiological chaos in its wake.

A surgeon from the 1800s, asked to perform an appendicectomy in my operating theatre, would be startled by the changes in culture and technology, as well as by the improvements in patient survival. But they would still recognize the fundamentals of the operation, because a great deal of twenty-first-century practice is based on nineteenth-century scientific principles. For several thousand years the leading cause of surgical mortality has been – and still is – infection. However, major breakthroughs and innovations in antisepsis and antibiotics in the nineteenth and twentieth centuries have transformed outcomes from surgical procedures. For my profession, 'scrubbing up' remains a rite of passage and a performance art. It offers time to think, time to mentally rehearse and, occasionally, time to panic. It's one of the most important acts of surgery: bacteria are the enemy, and this choreographed dance rids them from my hands. Most of my clinical practice has been spent in a state of war with microorganisms and, collectively, the medical profession has prescribed so many antibiotics, so recklessly, that bacteria are now resistant to these precious medicines. The frightening part is that we have no new medicines to replace them with.

Humankind has made extraordinary advances in all aspects of healthcare during this timeframe; for example, your chance of surviving cancer has increased by 50 per cent in just fifty years. But it hasn't solved all of our problems, and today society struggles to cope with a startling rise in the diseases and disorders of progress, many of which begin with the letter 'A', such as asthma,[3] allergies,[4] autoimmune diseases[5] and even autism.[6] The pattern of growth in these disorders across populations varies according to factors like our postcode, our wealth and access to increasingly sophisticated detection strategies. However, the steady and undeniable climb in the worldwide prevalence of these conditions can't be explained away by genetic variance or the shifting goalposts of diagnostic definitions. Many of these conditions disproportionately affect the young and those at extremes of age: in 2019, an estimated 38.2 million children under the age of five were overweight or obese and this number is steadily rising, meaning that when they become adults they will carry a disproportionate burden of diabetes and cardiovascular disease, the number-one cause of death worldwide.[7] Generation Z can also expect to have a risk of bowel cancer four times that of someone born in the 1960s.[8] Alzheimer's and dementia prevalence is climbing so fast in our rapidly ageing population that, by 2050, 151 million people worldwide will not be able to recognize their own families.[9]

Alarmingly, the majority of drugs we have to treat these conditions don't work in the majority of people, and the pharma pipelines that are supposed to be creating new treatments for our aching bodies are running perilously dry. As a result, my profession has had to adapt; we haven't only created new procedures – we've developed new surgical specialities. Bariatric surgery for weight loss, for example, didn't exist as a sub-speciality when I qualified; today, more than 250,000 weight-loss surgeries are performed each year in the US alone.

Twenty-first-century living is causing our airways to close, our skin to flake, our joints to swell, our guts to bleed, our arteries to clot and our brains to seize up. The global pandemic of non-infectious diseases is, arguably, a greater threat to humanity than that caused by any communicable disease. The healthcare systems set up to treat these diseases are increasingly unsustainable. If we are to meaningfully understand how and why we've reached this crisis – and how to solve it – we urgently need to reappraise our relationship with our microbes.

Throughout my career as a surgeon and clinical scientist I have become increasingly interested in one question: if the microbes that live inside us are so bad for us, why are they there to begin with? Why have so many of them chosen us as their hosts, and why have we offered ourselves so readily to them?

While some of those microbes might be harming or even killing us, what about the rest? Could they be helping us heal and grow, or even think and feel? Could our microbes be in conversation with our immune system – and explain why diseases related to it are increasingly common? Or why we get cancer, what the appendix does or even why some drugs work and others don't? For the best part of two decades I have sampled the microbiomes of generous patients undergoing surgery in theatre number three, in my quest to find answers to these questions. I've also followed the work of colleagues across the world and we've shared our findings and ideas. The answers have been surprising, beyond what we could have imagined.

Today, I am driven by the idea that the rise in the collection of disabling chronic diseases has been caused, over just eight decades, by the radical disruption to the colonies of microorganisms that live in and around us. And that, in our quest to cure the world of infectious diseases, we've inadvertently created a new pandemic of non-infectious ones. This book is my way of sharing my own journey to these conclusions, and my hope is that this will start a conversation about how we can provide a different type of medicine. The true promise of microbiome science is in disease prevention.

In March 2020 I stood in the same operating theatre and watched young, previously healthy patients being ventilated for COVID-19, the disease caused by the SARS-CoV-2 virus. They were being cared for in an operating room because we had run out of space in the intensive care unit and there was simply nowhere else to put them.

The surgeon's role was to help turn these patients on to their fronts while they lay in their medically induced comas. The aim was to redistribute the fluid in the lungs and help them breathe. It was grim work. The hospital was being overrun, and during the following weeks we sweated through our painfully inadequate PPE as more than 400 patients and members of our staff died from the coronavirus. Thousands of others suffered needlessly under harrowing conditions of enforced isolation while the government hosted parties and drank. At the time of writing, more than 670 million cases and 6.8 million deaths from the SARS-CoV-2 pathogen have been reported worldwide, reinforcing many of our deepest-held fears about the lethality of microbes. The pandemic reignited the war on bugs, a war that started with germ theory 150 years ago and mutated into a toxic mysophobia (an irrational fear of germs and contamination).

The microscopic life forms that preside over our health and wellness are increasingly frustrated with the mistreatment inflicted upon them by hyperglobalization – and they have a formidable molecular arsenal with which they are demonstrating their displeasure. The result is that although we are living longer than ever before in history, we are not living happier. Medicine doesn't have all the answers for this paradox. In response, some of my patients are now returning to ancient strategies for the treatment of their modern maladies . . .

Introduction: Good Sh*t

How lucky you English are to find the toilet so amusing. For us, it is a mundane and functional item. For you, the basis of an entire culture.

Baron von Richthofen, *Blackadder Goes Forth*

After two decades of surgical training, I specialized in the treatment of bowel cancer. I spend large parts of my working day talking to patients about their bowel habits, and many of them want to talk about little else. As a nation, we British are obsessed with our gut function, largely because it has never been unhealthier. There is also a deeper, more fundamental fascination with the digestive system; the colon is a national source of comedy that has kept us going through every crisis since the beginning of time. 'Shit' is a crucial and ubiquitous word that serves as a noun, a verb and an adjective, propping up the entire English language. This wondrous word is simultaneously a profanity and a term used to denote an item of high quality, and it is liberally sprinkled into the daily chatter of our lives and even into my operating room. 'Shit! What is that?'

The sense of revulsion we feel when we're faced with human excrement (or even just the thought of it) is, in part, a response to the way it looks and smells. The brown stain is caused by a pigment called stercobilinogen, a by-product of the gut bacterial metabolism of blood, specifically haem, the iron-containing molecule that binds oxygen in haemoglobin, and the foul stink is caused by the chemical products of intestinal fermentation. But that revulsion is also a psychological reflex, ingrained by potty training and social stigma. This aversion is an important safety mechanism: hand-washing and sewer systems prevent the spread of diseases that have killed millions.

But what if I told you that faeces was not toxic waste, and that it contained the secret to human health?

Would you eat it, if your life depended on it?

What if it was rebranded as a faecal microbiota transplant (FMT) or, more accurately, a faecal milkshake given through a tube that passes through the nose into the stomach? You could even take it in the form of a capsule – or 'crapsule' – if you wanted.

To help persuade you this might not be such a terrible idea, I'll tell you the tale of a patient. Raymond had driven the number-seven bus between Oxford Circus and East Acton from the age of twenty until taking early retirement in his mid-forties. After Ray's brother died all too young of a heart attack, Ray learned that he, too, had heart problems, which he was told were genetic. He gave up his job as a bus driver, on his doctor's orders, and used his retirement to learn about computers.

Like you and me, and everyone else on the planet, Ray was a host to several trillion microbes – the name for all living organisms that can only be seen under a microscope – that lived in and on his body. From our first breath to our last, and even beyond, microbes are our ever-present companions. While they take up residence in any number of places in our bodies, they're especially keen on – and abundant in – the various cavities and niches found in our gut. The 'gut microbiome' is the name we've given to describe not only the wildly diverse collection of microbes that live there, but also what happens when they interact with each other and with our bodies. In other words, it's an ecosystem made up of trillions of microbial life forms going about their business inside us, as we go about ours. In the last two decades, scientists from across the world have started to leverage the new science of the microbiome to transform how we conceive of human health.

Throughout his life Ray's gut microbiome had changed with him. As a child it helped his brain develop, it educated his immune system and it sustained and nourished him into adulthood. But as time passed, his genes and his gut microbiome began to do battle, causing multiple chronic diseases. His doctor prescribed an ever-longer list of medicines. Eventually Ray developed a type of leukaemia that left him profoundly frail. He had to rely on his wife Heather, a nurse at London's Great Ormond Street Hospital, for care.

Ray and Heather navigated his various health issues together until the day pneumonia struck. Ray became seriously unwell. When even breathing became difficult, he was admitted to St Mary's Hospital in Paddington, where he was treated with intravenous antibiotics. Without the drugs he would have died; with them, the infection was treated and he was discharged within the week. However, it was at this point that a terrible antibiotic firestorm started in his gut.

Imagine the worst-possible diarrhoea: opening your bowels more than ten times a day, incapacitating nausea, plus severe cramping pain in your abdomen, depriving you of sleep. Now imagine that you are frail, and your heart is working at 40 per cent of its normal function and your lungs are full of fluid. You can't breathe. Arthritis means you can't get to the loo in time. You are cold, clammy and profoundly dehydrated but can't drink enough to satisfy your thirst. You are soiled, but too close to death to care.

Three days after he had returned home, Ray's son called an ambulance. Ray was readmitted to St Mary's critically unwell and was soon diagnosed with *Clostridium difficile* infection (officially, this bacteria has now been renamed *Clostridiodes*). A 'hospital-acquired infection', this disease is a complication of twentieth-century medicine and an unintended consequence of Alexander Fleming's discovery of penicillin, the first effective mass-produced antibiotic, in 1928. It is a global problem that afflicts 500,000 people in the United States each year and it kills 29,000 of them.[1]

You can think of *C. diff* as a radicalized sleeper cell that waits dormant in the gut for its instructions from its antibiotic masters. In Ray's case, the antibiotic treatment destroyed the indigenous bacterial community and triggered an intestinal *C. diff* insurgency that escalated into systemic organ failure.

C. diff debilitates its host organism – the human being – by generating two toxins (enterotoxin A and cytotoxin B) that cause inflammation and destroys the lining of the gut. Inspecting the gut with a flexible camera passed into the bowel reveals an apocalyptic scene: massive destruction of the colonic architecture, which is left raw and bleeding. The particular strain of *C. diff* in Raymond's gut had engaged in an aggressive campaign of molecular warfare. His personal gut-microbe collection, carefully and uniquely developed since his birth and fuelled during lunches on the number-seven bus, was gone. His intestine was failing and he was dying.

The doctors at St Mary's quickly diagnosed his condition by looking for the toxins that *C. diff* makes in his faeces, and he was treated with yet more antibiotics. This seems counter-intuitive, but it is in accordance with best practice. Bacteria don't simply roll over and die, though. *C. diff* has a trick up its sleeve, which is to produce antibiotic-resistant spores that wait to germinate, biding their time. Raymond was given an antibiotic drug called vancomycin. Many patients will respond to vancomycin, but about one-quarter will relapse. And in those who do, 45 per cent will have a second relapse. These are the patients who typically benefit from FMT, or the 'good shit'.

Dr Ben Mullish, a clinical scientist at Imperial College London, was running a trial of FMT in patients with *C. diff* infections. Ray was so unwell that Dr Mullish offered him the treatment. Heather understood that there are good and bad bugs and advised her husband to go ahead with it, but Ray was not having it. The idea of taking another human's faeces was just too much for him, and he refused. Three days later, when he had deteriorated further despite the vancomycin, there was no other choice. Dr Mullish gained Ray's consent for the trial and set to work on preparing the transplant.

The logistics of preparing an FMT should not be underestimated. Faecal donors have to be found – harder than you might think. Most of us are squeamish about pooing in pots, and we struggle to do it on demand. Some studies use friends and families, others use members of staff, volunteers or 'pooled' samples taken from lots of donors mixed together. The complexity and demand for faecal transplants have spawned an entire industry, and FMT can now be purchased frozen from biobanks. Just as with any organ donation, donors have to be carefully screened to make sure they don't harbour transmittable diseases or parasites. Potential stool donors undergo a rigorous screening questionnaire, medical interview and examination, followed by blood- and stool-testing. Then there are the practicalities. Fresh samples must ideally be acquired within a short time from delivery, diluted with sterile saline, stirred, strained and then poured into a sterile bottle. Dr Mullish's job can at times be less than glamorous.

Once the faecal cocktail was mixed (shaken, not stirred), the transplant was administered to Ray during a colonoscopy. This procedure involves a flexible telescope that is passed into the colon through the bottom, and the bowel is gently coated in the soothing balm of microbes, which are passed through the colonoscope using a syringe.

While an FMT might be a new idea to many of us today, the medical practice of faecal transplant is ancient, and it has been drunk as 'yellow soup' since the fourth century AD for the treatment of infective diarrhoea. In 1958 an innovative surgeon, Dr Ben Eiseman, administered faecal enemas to his patients in Denver, Colorado with severe recurrent *C. diff* infections. It was remarkably effective, but like all important medical discoveries, this intervention was largely ignored at the time of its first report. More than half a century later, the Dutch gastroenterologist Josbert Keller and his team at the Amsterdam Medical Centre randomized patients with recurrent *C. diff* into three groups. The first group received vancomycin, a wash-out of the colon using a strong laxative, and a faecal transplant. The second had vancomycin and the colonic wash-out, and the third just received vancomycin. The FMT group did so much better than the other two groups that the study had to be stopped early, as it was deemed unethical to continue; 93 per cent of patients who had an FMT got better, compared to only 31 per cent with the vancomycin alone, and 23 per cent with the vancomycin and wash-out.[2]

The race to discover how FMT works is now on. We do know it restores the metabolism of bile (a green digestive fluid made in the liver and stored in the gall bladder), which is co-metabolized by bacteria, and this in turn blocks the germination of *C. difficile* and controls the infectious disease. It is also probable that a process of 'bioremediation' occurs, where the donor microorganisms consume and break down toxins that exist in the recipient's gut. However, there are trillions of organisms producing an infinite number of bioactive molecules, and each disease has a discrete microbiome. Therefore, it may really be a bit like turning a computer 'on and off' again; it's a complete reset of the gut's immunology software.

It's also becoming clear that samples from some donors are much more effective than those from others. These are known as 'super donors' and their faeces seems to contain a magical ingredient that makes it particularly effective. Before you don your cape and mask and rush down to your local faecal donation centre, you should also know that we don't understand why this happens, or whose poo will be most effective. Regardless, FMT is now being investigated with varying degrees of success in hundreds of trials across the globe. These include trials for inflammatory bowel disease, irritable bowel syndrome, obesity, acute malnutrition, diabetes, arthritis, hepatic encephalopathy (decline in brain function with severe liver disease), liver transplants, skin cancer, autoimmune diseases, Alzheimer's, neurodevelopmental conditions, bipolar disorder, hair loss, depression, neurodegenerative diseases and recurrent urinary-tract infections, to name but a few.

Some of these studies are extremely encouraging and offer treatments where few effective medical therapies exist. For example, FMT appears to be a promising treatment for irritable bowel syndrome, and a recent study suggests that its benefit can last for many years: 125 patients were randomly assigned to receive either 30g or 60g of faeces from the same donor or a placebo transplant containing their own faeces. Researchers not only found that the FMT improved the symptoms, but there was a lasting benefit three years after it was given.[3] The 60g group had a 71.8 per cent reduction in the severity of their symptoms compared to 27 per cent in the placebo group, with no long-term side-effects.

Unlike *C. diff*, where there is a dramatic and acute clinical change caused by a defined pathogen, the impact of FMT is less clear in chronic disease states, where it hasn't yet been proven that bacteria are the cause, or where we haven't defined exactly how the disease develops. As a result, at the time of writing in the UK and the US, recurrent *C. diff* is the only clinical condition for which regulatory bodies have approved the use of FMT. The bottom line (pun firmly intended) is that we don't understand how FMT works or its long-term risks well enough to start using it more widely in clinical practice. Nevertheless, because word of its incredible potential is spreading, there's a worrying growth in online enthusiasts offering back-street FMT 'cures'. I hope it goes without saying: please don't try this at home.

Ray's response to his FMT treatment was just like that in the reported literature. Within three days of receiving the microbiota transplant he was out of bed. Heather described it as a miracle.

If I've spent longer than is entirely comfortable talking about faeces, that's because FMT is a starting point for understanding the importance of the gut microbiome to human health. The extraordinary benefit of FMT in some patients has opened the clinical world to the idea that our microbes may have an important role in the causation and treatment of diseases where their involvement runs contrary to medical science.

The scale of the task is immense. The bacteria in the gut alone weigh close to 1.5 kg, they're made up of about 100,000,000,000,000 bacterial cells (that's one hundred trillion) – equivalent in number to the total number of cells that make up the human body – and they speak millions of different molecular languages.

Another major challenge in studying the microbiome is its physical distribution. The microbiome is dispersed across our bodies in different niches, each with varying total abundances of microbes. Being clear about our anatomical definitions is important. The gut is a long tube and it starts with the mouth, where the total abundances of bacteria are about 10^{11} cells (that is a hundred billion). It then passes through the gastric stomach (10^7), the duodenum (10^7), then through 5–6m of small bowel (jejunum 10^7 and ileum 10^{11}),

1.5m of colon (10^{14} or 100 trillion bacteria) and, finally, the rectum and the bottom.

The colon is a big fermenting engine that breaks down the more complex nutrients that are not absorbed by the small bowel. So when I say 'gut', I really mean the colon, as this is where most microbes exist, unless I name another specific region. Because the small bowel is carpeted with millions of microscopic finger-like projections called villi, it has a total surface area of 32m². The microbes that live at the tips of the villi billowing in the intestinal currents are different from those that live at their base, and in turn these are different from those that live within faeces. It's a bustling, chaotic, crowded and vibrant scene teeming with life. We are only just beginning to map all of the microbial life in this vast ecosystem – and to understand how it connects us to the world around us.

In this backward world, shit has become a therapy, used to replenish our delicate internal ecosystems, which are being lost as quickly as they are being discovered. Even with the impressive advances in biology, metagenomics and bioinformatics (computational biology), we might not be able to count and name all of the beneficial microorganisms that live inside us before they die out, mutate or evolve into something very different.

This book is not a history of faecal transplantation. It is a story about how we are deciphering the molecular language of the human microbiome, one of the great challenges of modern medicine. Faecal transplantation is a critical and fascinating tool that is being used to unlock these secrets.

Given that the microbiome influences almost every aspect of our lives, this story will be told in three parts. The first part takes us on a journey back in time to the birth of our planet and explores how microbes have shaped our biology, and how this profound evolutionary partnership with our microbiome connects us to our environment today. It tells the story of the microbial universe within us from the microbial big bang of our birth when we are first colonized by massive blooms of microscopic life forms, and how our microbes go on to shape our immune system, our sexual behaviour, our brain, our moods and feelings, our childhood and our death. It explains why our understanding of health is still defined by the breakthroughs in microbiology of the nineteenth and twentieth centuries, and why in the new world of the microbiome these frameworks are no longer fit for purpose.

The second half of this story explains how environmental forces outside our bodies are in constant communication with our microbes. It's a conversation that has changed dramatically in the past century or so, becoming increasingly fraught and stressful. Where we live, the air we breathe, the food we eat, the medicines we take – for better or worse, all of these factors influence the make-up of our microbiome. It's a global story of biopolitics, conflict, antibiotics and the medicines that have so profoundly damaged our microbiome.

The final part of this tale explains how the microbiome is shaping our future health. As big tech and big pharma become increasingly attuned to – and invested in – the extraordinary potential of the microbiome, there has never been a greater need to understand the microbes that live inside us.

We're only just starting to explain how, when and where the microbiome defines our risk of disease, but also how, when and where it sustains human wellness and happiness. The human microbiome represents the most important new therapeutic target that we have for treating the greatest threats to human life in the twenty-first century and for preventing future pandemics of pathogens. This was not only important for Ray; it is critical for all of us: without a stable and diverse microbiome, we may well lose our minds.

This book has been written in operating theatre number three during a global infectious-disease pandemic in order to spread a simple message: *microbes are not the enemy.*

PART ONE

The Microbiome

1. A Library of the Known Microbial Universe

The York Gospels are one of only a small collection of ancient Gospel books pre-dating 1066 to have survived the Reformation. They were written around the turn of the first millennium, in the scriptorium of St Augustine's monastery in Canterbury, and were brought to York by Archbishop Wulfstan around AD 1020. Genetic sequencing of the microbes that live on the York Gospels has revealed that the books contain populations of ancient bacteria that reside within their pages, and some of these bacteria originated from the skin of the Anglo-Saxon monks who wrote and read these religious texts. They also contain the destructive *Saccharopolyspora* genus, which causes a measles-like spotting of parchment that threatens this priceless historical artefact.[1]

As you turn the pages of this book, many hundreds of microbes – the name for all living organisms too small to be visible to the naked eye – will be exchanged between your skin and the paper of each page. The bacteria, fungi or viruses you share will be left as indelible physical markers of your interaction with it, pressed like dormant microscopic flowers. This phenomenon is the same for all the books that you have ever read. The quantity and diversity of the microscopic life forms found on the pages of this particular book will ultimately correlate with its popularity and how many times it is opened, folded, dropped, shared or abandoned. Not all books are treated equally, and as a result they harbour varying combinations of microbes that are typically found on human hands and bodies; Bibles are vulnerable to devotional kissing and may contain oral bacteria, while academic texts are full of sweat, blood and tears. Ancient books are bound in animal skins, bringing with them DNA and species of bacteria from different animal types, while digital books and tablets harbour their own collection of microorganisms that can be liberally shared with anyone who chooses to swipe a finger across their screens.[2]

Using tools such as microscopy, culturing and chemical dyes, skilled eighteenth- and nineteenth-century scientists discovered that our hands have two dominant populations of bacteria that could be responsible for contaminating a book. 'Resident' microorganisms typically live on the surface or just under the epidermis of the skin, in stable colonies that predictably reproduce. 'Transient' microbiota only colonize the superficial layers of the skin and only occasionally multiply. These are the bugs you remove when you wash your hands, and they are the usual culprits that spread infection or mark the pages of a precious manuscript.

While classical microbiology has transformed our understanding of how microbes known as pathogens cause infectious diseases, it doesn't tell us how intricate communities of microbes from different species or kingdoms work together to cause disease or maintain our health. In other words, it doesn't provide information about how bacteria, fungi and viruses communicate *among themselves* or *with you* to maintain healthy skin, or how the dynamics of these populations change under shifting environmental pressures of your daily life, like using soap, wearing gloves or stroking a pet. Collectively, we spent US $2.95 billion in 2022 on hand sanitizer and, although this reduces rates of pathogen transmission, it is fundamentally changing the health of our skin microbiome in ways that are not well defined or understood.

Microbiology makes way for microbiome science

If every piece of paper has its own microscopic living history, providing an indirect record of human behaviour and clues about our health, then each book also has its own microbiome: a living ecosystem defined by the genes from combinations of bacteria, yeasts and viruses and the environmental conditions needed to maintain their existence. Each page of a book has its own tiny niche of microbial species, seeded from the hands of the person who read it, contaminated by microbes acquired during eating or touching and sampled from the environment of whichever part of the world the book was read in. Think about that the next time you pick up a novel at the airport or train station and displace it in a foreign land.

Microbiome science is a departure from classical microbiology. It wants to understand how communities of microbial organisms support human life or cause disease. By studying the genomes, habitats and environmental conditions of all living microorganisms residing within a niche over time, it provides a measure of the collective functions of these interacting communities. A microbiome scientist looks at a library of books and sees a vast ecological time-capsule, where the bacteria of skin from ancient hands and mouths can still be found on the covers and pages of texts, giving us clues about the people who wrote and read them. From this perspective, your bookshelf can be thought of as an ecosystem that's unique to you. The choice of titles has been shaped by your economic, social, ethnic and biological status and these same powerful forces also influence the microbiome dotted throughout your collection. The microbiome can be studied at the level of the library, shelf, book or page.

Humans, too, have a number of ecological niches, or microbiomes. These are distributed across our anatomical landscape and organs such as the lungs, skin and urogenital tract. The largest collection of microbes lives in our gut, where their quantity equals the total number of human cells found within the body. No corner of the body is sterile all the time, and, just as on the skin, the colonization by microbes of our most important organs can be transient or persistent. Our different microbiomes may also have varying functions, which will change depending on the environmental pressures they are placed under. In general, our symbiotic bacteria (bacteria that live in a close mutualistic relationship with us) are important for our health, but defining quite how important isn't straightforward. We know we can live without some of them some of the time – we don't die every time we use the bathroom, and patients who lose their colons during surgery can live quite happily. But a true microbial niche implies a sophisticated co-dependence between microbe and human, in a mutually beneficial relationship that has developed over evolutionary timescales. This means that some of our organs have adapted to host these tiny passengers, and that our microbiomes influence both our risk of chronic diseases such as cancer and our well-being. However, because the human microbiome is so large, dynamic and variable between people, determining *who's there* and, perhaps more importantly, *what they are doing* is a huge challenge.

Microbial dark matter

In the early 1930s, while poring over his peers' recent observations of the Coma Cluster of galaxies, the Swiss astronomer Fritz Zwicky noted something strange. According to the measure of visible mass, galaxies were moving too quickly for the cluster of planets and stars to remain bound together. Or, to put it another way, they were not generating enough gravity to hold them together, and by his calculations these galaxies should have shattered. Zwicky suggested that a previously unidentified form of mass that he called *dunkle Materie* (dark matter) might explain this observation.

Dark matter consists of invisible, undetectable particles that influence everything within the cosmos. It explains how stars move within galaxies, how galaxies pull on each other, and how all that matter got clumped together in the first place. Dark matter is ubiquitous and essential for life; it is unseen but felt in every corner of the universe.

Its opposite counterpart is 'dark energy', which describes a force that appears to be driving the expansion of the universe. If Zwicky's dark matter holds the cosmic flesh in place like an invisible skeleton, then dark energy continually tries to pull it apart.

Dark matter and dark energy are not explained by what's known as the 'standard model', which defines how elementary subatomic particles make up all known matter. The standard model says nothing about the 95 per cent of the universe that physicists believe is not constructed from normal matter.

Modern medicine is similar in this respect: we have a comprehensive model that explains a large part of the known mammalian biological processes that we can see, and which we use to explain the majority of the diseases we experience. But the standard model of medicine fails to account for the majority of genomic data in the human body, which we can't yet see.

You can therefore think of these microscopic life forms within us as a sort of biological dark matter. We only have a superficial knowledge of how they promote health and bind together the spinning human galaxies of interconnected organs. These organisms also produce dark energy, a chemical and immunological force that sustains us – and that will ultimately destroy us, if we allow it to become unregulated. The human cosmos feels the effect of microbiome dark matter and the energy it produces every day. Because twentieth-century medicine couldn't see it or measure it, its influence on human health has until now been attributed to the components of our biology that can be determined by human genes and measured by X-rays, CT scans and blood tests.

From human to metahuman

Beyond the protection of important texts, a librarian's main job is to maintain order. This is a complex task, once achieved through an intricate and laborious process of manual indexing and labelling. That was until the late twentieth century, when computing transformed the speed and accuracy with which these tasks could be performed. The same processes have dramatically influenced how evolutionary biologists index all life on Earth.

In 1837, Charles Darwin scribbled out a picture of a tree, an illustration that he used to explain the evolutionary relationships between species. In this metaphor, all life on Earth grew from a single organism, and the major divisions (or branches) on his tree could be divided into animals, plants and 'protists'. Protists were microscopic organisms that couldn't be neatly packaged into the animal and plant divisions. This tree underwent many modifications, and by the 1990s it was assumed that all living organisms could be divided into just two main evolutionary branches or 'domains' of life: prokaryotes and eukaryotes.

A eukaryote is an organism with complex cells, or a single cell with complex structures that exist within it. In these cells the genetic material is organized into chromosomes, which are safely stored within the cell's nucleus. You are a eukaryote, but so is a fungus, a plant and a polar bear.

Prokaryotes are organisms that lack a cell nucleus, and this is the domain of life where bacteria live. You may think of fungi and bacteria as very similar inconsequential microscopic organisms, and it would seem logical that they are related. But an evolutionary biologist would argue that you have more in common with a fungus (your fellow eukaryote) than a bacterium does, because fungi and bacteria are from different domains of life.

Each of these domains divides again into kingdoms and, within each kingdom, there are six further major divisions: phylum, class, order, family, genus and species. Each describes a shared anatomical detail or physiological function, and the microbiome can be described at each of these divisions, before it branches into vast clades, or groups of species.

Perhaps the greatest biologist never to win a Nobel Prize was a man named Carl Woese. In 1977 this revolutionary made a discovery that would shift and expand our understanding of the evolution of the natural world, which would re-draw the Darwinian tree-of-life model.[3] His genius was to adopt a strategy of defining the similarity of species not on the basis of their morphology (anatomy), physiology or metabolism, but on their genetic code. Instead of using the newly discovered DNA, he chose a molecule called ribonucleic acid (RNA). One of RNA's many jobs is to make proteins and, to do this, specific types of RNA bind to a cellular machine called a ribosome. Woese and his team managed to elucidate the structure of ribosomes and work out that small sub-units of ribosomal RNA are highly conserved between species – an amazing feat when you consider the staggering number of ways the ribosome can be folded and created. Within a single bacterium called

Escherichia coli, for example, the potential variations in ribosomal structures are larger than the total number of elemental particles in the entire universe.[4]

Because comparatively small regions of the ribosomal RNA are discretely conserved between species, it can be quickly and affordably analysed to identify organisms based on their genomic signature. With this innovation, Carl Woese and his collaborator George Fox identified a third domain of ancient microscopic life, as distinct from bacteria as bacteria are different from plants, animals and you. They called this domain the archaea, and it comprises some of the most ancient microbes on the planet. Their varied and exotic characteristics allow them to survive in conditions that should be impossible, and some archaea are known as 'extremophiles'. For example, *Methanopyrus kandleri* can grow in temperatures up to 122°C and can survive in atmospheric pressures 200 times greater than those found on the surface of the Earth. You literally couldn't boil it or crush it to death. *Ignicoccus hospitalis* thrives on hydrothermal vents on the bottom of oceans, and other species of archaea can be found in acid mines and in caves that have never seen light.

Collectively, archaea play a key part in maintaining the biogeochemical health of the planet, and they are also an integral part of your microbiome.[5] These extremophiles are well adapted to live in and on us, and archaea have been found in the human gut, mouth, vagina and on the skin. Archaea such as *Haloferax massiliense* are able to live in high salt conditions and they help us metabolize a Western diet, while other species play an important role in removing heavy metals. They also help maintain intestinal health by producing the gas methane; these archaea are known as 'methanogens' and account for about 10 per cent of all anaerobic microscopic life forms that live in the gut. Methanogens in the intestines of our livestock, and those that live in our sewage systems, oceans and wetlands, are also shaping our climate, and they have contributed to a 0.50°C rise in our planet's temperature increase since pre-industrial times.

Carl Woese's work provided a 'metataxonomy', or a microbial ribosomal RNA gene inventory, from which all life on Earth could be indexed. This decoupled evolutionary biology from the science of biological organization, and it laid bare the extraordinary diversity of bacteria and archaea. For example, one recent phylogenomic analysis demonstrated that of the 135 described phyla across the three main kingdoms of life, 104 belong to bacteria, twenty-six exist within the domain of the archaea, and the eukaryotes – the domain where you, I and every other animal species reside – possess a paltry *five*. Estimates of the actual number of bacterial and archaeal species living on our planet now range from millions to billions.

If you're wondering where viruses fit into this system of organization, the answer is that they don't. Viruses are obligate cellular parasites that borrow the machinery they need to replicate and survive from their hosts, which means they are not technically 'living'. They exist instead within a separate 'empire' to other cellular life forms – one that's also defined by its incredible diversity.

By the end of the twentieth century the race was on to unlock the secrets of DNA. The human genome was finally mapped, at a cost of nearly US $3 billion, by the Human Genome Project (HGP) in 2004. At its most basic level, this discovered that a human is coded by approximately 20,000 'genes' composed of precise strings of base pairs that code for proteins made by RNA. The human genome is remarkably stable between people, and even between species. You and I share 99.9 per cent of the same genes, and in turn we share 98.9 per cent of these genes with our closest evolutionary relatives, such as the chimpanzee.

But the Human Genome Project forgot something crucial: its gene library was figuratively covered in microbial life forms. In this new 'metahuman' model of the internal biological universe, the human genome makes up less than 1 per cent of the genetic code that influences our health and well-being, and the remaining 99 per cent originates in the community of microbes that live within us, known as the 'metagenome'. Unlike the human genome, this is massively diverse between individuals. People from different families or parts of the world will share less than 10 per cent of species that reside in the gut, and even less with a chimpanzee.

Nevertheless, mapping the human genome turbocharged the

rate at which scientists are able to mine and interpret any genome from any eukaryote, prokaryote, archaeon or virus. As we'll go on to see, the consequence has been the discovery of the true complexity of both human and planetary microbial systems that govern our health.

From mapping the human genome to mapping the human microbiome

One of the pioneering explorers of microbial life is Craig Venter, the American biotechnologist and CEO of the company Ceres, which led the commercial attempt to map the human genome. In 2003, more than 170 years after Charles Darwin's voyage on HMS *Beagle* led to *On the Origin of Species*, Venter set sail on his own journey to the edges of the mapped scientific world aboard the *Sorcerer II*. His mission was to sample the microorganisms of the Sargasso Sea, which is found in the Atlantic Ocean right at the point where four circulating oceanic currents meet. A perfect spot for fishing for unidentified microbial life.

Using a metagenomic strategy, Venter found DNA from nearly 2,000 different microbial species in the Sargasso Sea, including 148 types of bacteria that were previously unknown.[6] The more of those organisms he and others mapped, the more of the minuscule world of bacteria came into focus, not just in the depths of the ocean, but deep within ourselves. His voyage began a much bigger odyssey – perhaps the most consequential one for our understanding of life on Earth since Darwin set forth from Devonport.

The term 'metagenomics' was first used in 1998 by Jo Handelsman, a scientist from Wisconsin, to describe the collective analysis of the gene content of a community of soil microbes. After the success of the Human Genome Project, microbiome prospectors in the early 2000s, such as Venter, began to coalesce around the idea that genomics could be used to sequence multiple organisms simultaneously – and metagenomics blew the doors off the microbiome world. But analysing whole genomes across hundreds or thousands of species all at once is a massive computational headache. And data storage is also no small challenge; if you were to store all of the metagenomic data from the gut of a single human on 256GB smartphones, you would need 400,000 of them.

Microbiome science needed new bioinformatics pipelines to be built to compute the volume and complexity of this new type of data. In 2007, the National Institutes of Health (NIH) launched the Human Microbiome Project: mathematical tools had to be invented from scratch that made interpreting and visualizing data manageable by thousands of scientists across the globe. From a mathematical perspective, it was the equivalent of sending a rocket to the moon, and the computing demands are so great that many of these calculations still need supercomputers.

Bacteria are the most studied class of microbes in the human microbiome, and the largest microbial niche lives in the gut. Just four of the 104 bacterial phyla predominate, namely the Firmicutes, Bacteroidetes, Actinobacteria and Proteobacteria, although there are others in smaller abundances. Around 2,000 species of bacteria have been identified in the gut – roughly equivalent to the number of species of trees found within the Amazon rainforest. But the average gut contains several hundred species of bacteria – which ones and how many are determined by our genes, diet, lifestyle and environment. Even in the world of metagenomics, it is still important to understand how individual species and strains of microbes function.

This field has also massively expanded to incorporate many other measurements of microbial molecular functions, like metabolism (metabonomics), proteins (proteomics), RNA transcripts (transcriptomics) and hundreds of other 'omics' sciences. All of these are being integrated into the great microbiome computational algorithm. The result of this work is that, for the first time, we can now not only study which microbes are present in an ecosystem but we can also measure what they are doing over time and how they are doing it. Microbiome science may have originated in oceanic and soil ecosystems, but it allowed us to imagine that the workings of our internal ecology can be deciphered to prevent disease and improve our health.

Microbial monsters

Because we know more about bacteria than any other type of microbe, they tend to dominate conversations about the human microbiome – but this doesn't mean they are always the most important players.

Microscopic eukaryotes, such as protozoa and parasitic worms known as helminths, may only make up about 0.1 per cent of the human gut microbiome, but they play an important part in the health of our internal ecosystem. Larger multicellular species of protists, with heads, mouths and guts that contain their own microbiomes, feast on bacteria and smaller prey in the blackness of the human Mariana Trench. Helminths come in a variety of unappetizing shapes and sizes that are guaranteed to stop you sleeping. Nematodes (roundworms), cestodes (tapeworms) and trematodes (flatworms or flukes) are carried by two billion people across the world. Parasite infestations are not, however, solely restricted to the gut of low- and middle-income countries.

The common protozoan parasite *Giardia duodenalis* can be identified in the stools of 2–5 per cent of apparently 'healthy' people living in industrialized countries like the United States.[7] These can therefore be thought of as 'healthy' components of the gut microbiome, which – as we shall see – play an important role in the maintenance of the gut's immune system.

Another of the main players in our gut are the fungi. Fungi are robust, and they can be found growing on nearly any substrate on Earth, from deep ocean sediments to the space between your toes. Just over a hundred species of fungi have been found living within the human mouth, and 260 species have been found within the gut,[8] making up less than 0.01–0.1 per cent of genes in human stool samples. Far fewer species live within the warm, wet conditions of the pungent human foot, but part of the reason fungi are so happy in this particular spot is because it gets sweaty – fungi need moisture to grow. On the scalp, the species *Malassezia furfur* (*Pityrosporum ovale*) causes dandruff, which is why shampoos that treat this condition typically contain antifungal agents.

In your gut, fungi (known as the mycobiome) are important for maintaining microbial community structure, metabolic function and immune-priming, although these interactions remain relatively unexplored.[9] For example, *Candida albicans* acts like a kind of vaccine, switching on the body's immune system so that it can produce antibodies to other more potentially harmful fungi that the body has yet to be exposed to. Gut fungi confer so many benefits on humans; some, such as *Saccharomyces boulardii*, are used as probiotics to prevent and treat pathogenic bacterial infections.

Fungi also produce a huge number of small molecules known as metabolites that are not considered to be necessary for their growth, which we refer to as secondary metabolites. These compounds have wide-ranging biological effects and include potent poisons known as mycotoxins. Perhaps the most widely known is the carcinogenic aflatoxin, produced by *Aspergillus flavus* and *A. parasiticus*. Some are sought out for medicinal use, such as the antibiotic penicillin, and others for recreational drug use, such as psilocybin, a psychedelic compound produced by over 200 species of fungi. If you consume magic mushrooms, you should know that your trip will be influenced by the bacteria in your gut, as intestinal bacteria can co-metabolize these agents on your behalf. Even psychonauts need a healthy gut.[10]

The fungi in our gut need bacteria to survive, and vice versa – but this relationship is far from simple. For instance, strains of the bacterium *Escherichia coli* suppress the growth of the opportunistic fungal pathogen *Candida albicans*, which also colonizes the gut of healthy humans. The relationship between bacteria and fungi or other eukaryotes exists through a process of competition for essential nutrients, symbiotic cross-feeding and even through hunting and consumption. Eukaryotes can have complicated sex lives, and this also leads to a lot of microdrama and co-dependence. As a result, fungi influence how the overall assembly of the microbiome takes shape in early life,[11] and it's in the interest of your gut and your health to keep the relationship between the different components of the microbiome in check.

The viral puppet master

As the largest contributor to our intestinal dark matter, viruses dramatically shape the microbial universe and influence our health through a continuous process of competition and evolutionary pressure. The total number of viruses found on planet Earth is so large that it defies belief. If all the viruses on Earth were laid end to end, they would stretch for 100 million light years, which is approximately 330 times around our galaxy.[12, 13]

No matter how hard you might try, it's impossible to avoid interaction with viruses. Each day you breathe in more than 100 million of these zombie microbes, and the lung has its own distinct virome. Within the gut, the total numbers of cells of the bacterial microbiome and the virome are pretty evenly matched.[14] But there are over 142,000 species of viruses currently thought to exist in the human bowel,[15] massively outgunning the 2,000 species of bacteria described in the same human ecosystem by roughly 56:1.

As the world is now only too aware, the *raison d'être* of a virus is reproduction. A virus will invade a cell, insert its genetic material into its host's and then use this to create tens of thousands of copies of itself. Its end goal is to inactivate or kill the cell (through a form of programmed cell death known as apoptosis) so that it can then exit the cell.

It's not only humans who are vulnerable to viruses; so are all prokaryotic life forms. Viruses that replicate within bacteria are called bacteriophages, or 'phage'. Phage come in a wide variety of genome sizes and compositions and there are an estimated 100 million species in existence.[16] The most studied of these look like alien lunar landers and this is no accident; they're designed to land on the moon-like surfaces of bacteria and then pin themselves firmly in place using their tail fibres, before injecting their genomic payload.[17, 18] If a lytic phase where a bacterium bursts is triggered, the bacterium will then explode and die, releasing the parasite's young into the wild to begin the whole cycle again. This is so important to us, because this is the process through which bacterial populations are managed and by which viruses have edited their genes into ours over millions of years. Our knowledge of phage is minimal and this presents a huge challenge; over 90 per cent of viral sequences identified through metagenomic analysis have no known reference points in our genomic databases. The vast majority of the human viral microbiome remains unmapped, and we simply don't know who's there.

Half of the bacterial population worldwide is destroyed every forty-eight hours by bacteriophage.[19] That is a lot of bacteria. But as they go about their parasitic endeavours, splicing their genes into their host's code, bacteriophage also encourage the sharing of genes between bacteria. The resulting gene-flow networks lead to changes across entire ecosystems. Phage found in the oceans, for example, have been referred to as puppet masters because of their key role in shaping oceanic biogeochemistry.[20]

If these viruses can be puppet masters of entire oceans, what do they get up to inside our bodies? Viruses directly regulate and interact with the immune system of the gut and influence its ability to function. But they also allow bacteria to share genes and to develop their own survival advantage or genetic memory that protects them against future parasite infestations. For example, phage are deeply embedded in the genomes of bacteria and archaea, and their short segments of viral DNA have been given the catchy name of 'Clustered Regularly Interspaced Short Palindromic Repeats', better known as CRISPR. The discovery that bacteria use CRISPR-associated proteins to splice in and out viral genes gave birth to what is now known as gene editing. This technology will transform medicine and provide cures for genetically inherited conditions such as cystic fibrosis, haemophilia and sickle cell disease. Its applications are limitless, and we owe it all to a primitive bacterial immune system.

The human microbial universe

The father of microbiology, Antoine van Leeuwenhoek, made the first discovery of microscopic life, which he referred to as 'animalcules' in a letter to the Royal Society in 1676. In keeping with all truly radical scientific discoveries, the leading scientists of the time immediately dismissed its significance. Although they slowly changed their minds, it was another 300 years before Carl Woese was able to complete the main divisions of microbial life, which are still challenged to this day. Van Leeuwenhoek's observations were made without a formal scientific education, by searching for life in water with a self-made, single-lensed microscope. He could never have known that he was really the first man to look into microscopic space.

Antoine van Leeuwenhoek and the astronomer Fritz Zwicky had a lot in common; their observations of the physical and natural world existed at scales their colleagues could not yet imagine. Although they were staring into different constellations, what they saw was profoundly disruptive to the way we conceive of our place within the biological and physical universe. And both scientists were restricted by the limitations of the machinery and tools at their disposal in the study of their subject.

Billions of galaxies of bacteria, archaea, viruses, phage, fungi, helminths and protozoa circle around the life-giving heat of a single human star. Some of these galaxies can be found living across entire organs, such as our skin, while others exist within microscopic niches – think of a crypt or gland in the gut. Some form within single human cells and burn brightly for only a short time before being extinguished by the immune system. Many niches stay with us for our whole life. But they're all intricately connected with each other, with you and with your environment. They're in competition for resources, and in this microscopic food pyramid, predators hunt smaller microbes for survival. The essential by-products and waste from metabolism are also consumed as part of a living network that spans the gut and the body.

As a society we've become disconnected from our internal ecology: it exists in body cavities that we can't see and at a scale we can't imagine. But this living universe is not dark; it's vivid and brightly coloured, and we feel its presence every day. We may now have a library with which we can create order from human microbial ecosystems, but we still lack an understanding of some of the absolute fundamentals of the human microbiome. Human microbiomes are dynamic communities that exhibit tremendous compositional and functional heterogeneity over time and space. A simple taxonomy of the human microbiome is therefore inadequate for describing what a 'normal' microbiome is. A description of a healthy microbiome must incorporate a form of 'quantum biology' which explains the importance of the microbes we can't yet see and the forces that govern their interaction with each other and with us. This new mathematical model must also be able to explain how the microbiome has evolved across the books and libraries of human existence.

It is time to turn the page.

2. Midlife Crisis

As I write this, I am forty-five years old, and the planet is about 4.54 billion years old, which must mean we are both having a midlife crisis. So far, the major consequence of my transition into middle age has been a counterproductive tendency towards nostalgia and a sorry descent into Lycra and road cycling. The planet's existential crisis, however, threatens all life on Earth and I grudgingly concede that it will be of more interest to you than mine will.

Human and planetary bodies have a dynamic relationship with their microbes, and this exists from the moment of conception. Mammalian and geological microbial systems are also deeply connected, and to understand why microbes are so important to our health and our future we need to travel back in time, to the very beginning of life on Earth, and examine this relationship through an evolutionary lens.

The story of our planet began properly when water formed on the Earth roughly 4.28 billion years ago; the primordial oceans were not blue, but rusty green from the iron sediment dissolved in the salty water. In an evolutionary instant, the first life forms appeared as primitive microscopic extremophiles coalescing around hydrothermal vents forming along fissures on the ocean floor. These geological chimneys allowed the life-giving heat from the planet's core to seep into the ocean and nourish the planetary infant. Just like a newborn human, the seeding of these microbes played a vital role in determining our planet's future.

The earliest direct evidence of life is thought to come from fossils of microbes found in 3.465-billion-year-old rocks in the sparsely populated, very hot and dusty region of the Pilbara, in Western Australia. We have two basic theories to explain how they got there. The most widely accepted theory, known as abiogenesis, suggests that self-replicating life forms developed through the chemistry of carbon and other key elements. Nucleic acids were critical for the formation of DNA and RNA; Woese's hypothesis proposed that before living cells existed, the genetic code for life was contained in RNA. Amino acids were built into proteins that could perform molecular functions, lipids and fats were used for building cell membranes, and carbohydrates and sugars were used for energy. From these building blocks, microbes became the first living cells on the planet.

The second theory is known as panspermia, and it posits that life exists throughout the universe, and that it is distributed between planets by space dust and meteors.[1] Before discounting the panspermia theory as total nonsense, consider this: it's been known for many years that the spores of *Bacillus subtilis* can survive in space.[2] And between 2015 and 2018 a team of researchers from twenty-six universities and institutions in Japan took part in the Tanpopo mission – an orbital astrobiology experiment that took place at the Exposed Facility on the exterior of the Kibo module of the International Space Station. Remarkably, a bacterium from earth called *Deinococcus radiodurans* was also able to survive for up to three years outside the space station in the freezing, hostile vacuum. Fungi and bacteria have also been found growing inside the space station and, intriguingly, three of the bacteria are from an entirely new species named *Methylobacterium ajmalii*. If these bacteria can grow in the cold, dark silence of space, then life on Mars – a nice warm planet with lots of water – is a piece of cake.

In California, the Jet Propulsion Laboratory's astrobiology division is currently searching for evidence of microscopic life forms in our solar system and beyond. They're also concerned with protecting the planet from microscopic alien interlopers that could hitch a ride back to Earth on Martian rock collected by the Perseverance rover. These rock samples will be fired into space and back to our planet as part of the Mars Sample Return Mission in the 2030s. If there are microbes in those samples, we will have to rethink everything we know about the origins of life on this planet. And as we shall see, the microbiome is not only challenging some of our most fundamental beliefs about the birth of life on Earth. It is also forcing us to reconsider the importance of microbial life forms to human fertility, conception and growth.

The planetary microbiome

Wherever life came from, it has been very persistent. From the moment the seed of the tree of life was planted, it slowly and inexorably began to send roots deep down into the oceans and the rock. A form of blue-green algae known as cyanobacteria quickly began making food from the sunlight and started releasing oxygen into the environment. With photosynthesis, the world was slowly oxygenated, turning the oceans a light-pink colour as iron was oxidized and the cyanobacteria bloomed to become one of the most prolific life forms on the planet.

Cyanobacteria are found within the human gut today and continue to contribute to human and planetary health. For example, laminated layers of these blue and green algae form mats, which collect at the lowest ecological depths of freshwater lakes in the Antarctic Peninsula. Climate change means that these algae, known as 'benthic mats', have started to produce toxins that are destroying the ecology of these critical sources of aquatic biodiversity. If this can happen in aquatic microbial ecosystems, then it can happen in those that exist within us and the consequences for our health are just as serious.

On the fledgling planet, oxygen levels continued to rise, thanks to microbes, until 2.3 billion years ago, when the blue planet had a childish tantrum. It produced so much oxygen that it poisoned the majority of 'anaerobic' microbes for which this bounteous, odourless gas was highly toxic, causing what was likely the planet's first mass-extinction event. Today anaerobic bacteria can still be found everywhere and even predominate in our gut.

As the planet's microbial systems recovered from this toxic insult, 'aerobic' microscopic life forms adapted and developed tolerance to the gas, and new life forms also began to emerge. Eukaryotes, and their precious cellular cargoes of membrane-bound DNA, emerged approximately 2.1–1.6 billion years ago.[3] In 1967 the American evolutionary theorist and biologist Lynn Margulis suggested that this event happened when one cell, which was most probably an archaeon, got swallowed by another.[4] Like so many radical biological theories, it was initially rejected by the scientific mainstream. It is now widely accepted, and it is probable that through this process of endosymbiosis, DNA became packaged into nuclei and other cellular structures began to form. One of the most important of these was the mitochondria – microscopic intracellular machines that use oxygen to produce energy. Almost every human cell type (red blood cells are a notable exception) contains one of these marvels of cellular engineering, and they even have their own genome. Mitochondria are your own ancient internal nuclear reactor: without them your heart wouldn't beat, and you wouldn't think. We owe our whole existence to ancient microbes.

As the planet reached adolescence, it started changing shape and growing awkward appendages. The first came about 2.5 billion years ago in the form of a supercontinent called Columbia, and with this the terrestrial microbiome came into being. Algal scum formed on the land about 1.2 billion years ago, slowly evolving into land plants over the next 750 million years. Fungi dissolved rock into earth and began their love affair with these plants, forming a deep and lasting symbiosis. All of the major microbiome players were on the scene, and terrestrial microbiomes were making significant contributions to planetary health and the world reached a form of biological teenage independence.

Like a Russian doll, life on Earth continued to find ways to compartmentalize itself, and in the evolution of organs, skin most likely came first. In a bid to separate the outside world from within, sponges created a proto skin for this purpose.[5] Since microbes were already in existence, the first skin microbiomes must have then formed. At the end of the Ediacaran period (635–541 million years ago) several monumental evolutionary events took place, including the dawn of animal life. Ancient worm-like animals made up of a single tube started to grow in large numbers – these were collectively called cloudinomorphs. Amazingly, CT scans of their

500-million-year-old fossils suggest they had working guts.[6] This is now the earliest available fossil evidence for the emergence of the intestine, and it was the point at which animal life let in microscopic life to support its own existence. This deep endosymbiosis ultimately led to the gut microbiome – one of the most important events in the development of animals (not least because, without it, I wouldn't have a job).

The following Cambrian era saw an explosion of animal life forms that began to prioritize defence against predation, and this did not just mean shells, spines and armour. The primordial guts to support the rapid growth in size of these increasingly large marine and land animals not only had to digest other animals and plants, but they also had to safely manage the associated microbes and toxins that entered the gut. This meant that they had to develop a method of defence against microbes, which came in the form of an immune system.

The prototype immune system first had to solve a fundamental problem, which was that it needed to differentiate 'self' from the pathogens that might wish to cause the animal harm. The solution to this challenge came from microbes themselves. The basis of the immune system requires binding foreign molecules, or particles known as 'antigens', to specific receptors of the immune system known as 'antibodies'. This lock-and-key mechanism was so fiendishly complicated because there is an infinite number of antigens, which in turn require an infinite number of antibodies.

No one is precisely sure when viruses came into existence, but approximately 1.5 billion years ago they started to develop specific folds in their proteins called enzymes that allowed them to break into other life forms where they could replicate. They not only became specialists at breaking and entering, but viruses also became genetic mixologists, creating new combinations of genes through a process of gene cutting and pasting.

It was a virus infecting a bacterium that transferred what is known as a 'recombination-activating gene' that enabled the giant leap in antibody production to occur. The earliest evidence of primitive versions of this 'adaptive' immune system arose approximately 500 million years ago in jawed fish.[7] Using this molecular machinery, they could suddenly programme the cells of the immune system, called lymphocytes, to produce antibodies in an infinite number of patterns, and this in turn provided a powerful survival advantage. This ultimately led to the evolution of IgM, the most ancient of our antibodies that you and I still rely on today.

Perhaps more importantly, the great DNA swap was now in full effect, and viruses, archaea and bacteria began weaving their DNA into the genomic codes of both prokaryote and eukaryote species. This is known as the 'horizontal gene transfer' of genetic code, which describes how genes have jumped between species to help evolve new life forms. Fragments of ancient viral genes are entombed in our DNA like ancient molecular fossils, and they are so common that they make up about 8 per cent of our total genome. Although we are not sure of most of their functions, we can use them to rewind the molecular clock to trace their origins. Many of these genes point to global pandemics that were shared across species, and which lasted for millions of years. Some of these pieces of viral code in our genome are hundreds of millions of years old and originated during the time of the dinosaurs.[8]

During these years of early planetary adulthood, the world also found a much more fun method for sharing genes. Until this point, asexual bacteria and other microscopic life forms had exclusively adopted methods that prioritized a fast rate of reproduction over foreplay; the name of the game was gene transfer at any cost. But when eukaryotes came along, they prioritized quality over quantity in their reproductive strategy. The birds and the bees might do it, but fish started doing it first, about 385 million years ago. Evidence for this comes from fossils of the ancient Scottish fish aptly named *Microbrachius dicki*. This chinless fish possessed large, bony L-shaped claspers used for transferring sperm from the male to the female.[9] Yes, sex was invented in Scotland.

Given that sex, when done correctly, takes a lot of time and effort, one may wonder why early animals deviated from the bacterial method and went to so much evolutionary trouble to adopt it. Many theories have been put forward for this, but the Red Queen hypothesis – named after Lewis Carroll's character from *Through the Looking Glass* – suggests that the point of sex is to fight disease, which (assuming you are practising safe sex) is another excellent reason to have lots of it. Or, to put it more scientifically, sex might give an advantage to individuals in rapidly changing environments, through a co-evolution between a host and its parasites.[10] Even if it is not the dominant reason for engaging in this form of reproduction, it is still highly likely that microbes formed part of the evolutionary reason as to why we have sex. Whatever it was, it certainly wasn't pleasure, which has been a modern addition to the process, and an orgasm is not something that the *Microbrachius dicki* got to enjoy.

The planet's land masses exist in a constant state of migration, slowly but surely locking microscopic life forms into discrete evolutionary niches. The last great supercontinent, known as the Pangaea, assembled 335 million years ago, and broke up into the seven continents that we know today at the end of the Triassic period around 200 million years ago. Dinosaurs inherited the Earth for a brief time during this period (at the rate we are going, ours will be even briefer) before dying out during the Triassic–Jurassic period 201.3 million years ago.

In Michael Crichton's classic 1990 novel *Jurassic Park*, scientists reconstruct the genome of a dinosaur from a drop of blood captured from a mosquito set in amber. Recently the evolutionary geneticist and biological anthropologist (and son of the scientist who inspired the book) Hendrik Poinar identified a 100-million-year-old mosquito preserved in amber, which contained developing reproductive cells of an archaic malarial parasite called *Paleohaemoproteus burmanicus*.[11] The decline of the dinosaurs actually took thousands of years, and it's likely that, with global warming on their side, pathogens such as this played a significant part in their demise through the spread of disease. The descendants of this mosquito are far more terrifying than *Tyrannosaurus rex*, as they've been responsible for the death of half of the humans who have ever lived.[12] There's an important lesson here, which is that global warming potentiates the spread of pathogens across the globe, and it should be no surprise to you that in 2020 more than 200 million people were infected with malaria, and this is increasingly happening in Europe and northern parts of the globe.

As the dinosaurs died away, mammals began to inherit the Earth and the first mammary glands appeared during the Triassic period about 200 million years ago.[13] Once again, reproduction helped to define our co-evolution with the microbiome. In their first incarnation, mammary glands provided moisture and antimicrobials to parchment-shelled eggs, before evolving into the role we know them for today: supplying sustenance to hungry offspring.[14] Not only did early milks contain antimicrobial complex sugars to help suppress pathogens, but the breast also provided a rich source of nutrition, which allowed the microbiome to bloom and infants to grow rapidly before being ejected from the nest.

Primates finally arrived on the scene about fifty-seven to eighty-five million years ago. The oldest fossil records of the chimpanzee–human last common ancestor are dated to the late Miocene period (about seven million years ago), although the earliest Neanderthal skulls are only 400,000 years old. *Homo sapiens* became the only surviving human species just 40,000 years ago, a species that possessed its own distinct microbiome, the culmination of more than four billion years of microbial evolution.

In July 2020, an aerobic microorganism was found in quasi-suspended state 76.2m below the seabed of the South Pacific Gyre, at what is referred to as 'the deadest spot in the ocean' because it has the fewest nutrients available to marine life. These aerobic microbes were 101.5 million years old and had survived by adjusting to a carbon- and nitrogen-based metabolism.[15] Incredibly, these organisms could be reanimated in the laboratory and woken from their ancient slumber. They undoubtedly hold important secrets that will help lessen the misery of human ageing.

The more important takeaway from this discovery is that some microbes on this planet have been asleep longer than primates have existed. We turned up to the Anthropocene ball at 23.58 hours and 43 seconds on the evolutionary clock, demanding that someone open the bar, not realizing that the microbes were running the whole establishment. They were already deeply embedded within our evolutionary soul, acting like puppet masters of our biology, and our futures were inexorably linked long before we walked the Earth.

The evolution of 'Homo holobiont'

The new biology of microbiomes challenges traditional evolutionary theory because it sees the microbiome as an extension of an animal's genome. And because microbes are the most abundant and ancient life forms on Earth, they've influenced the evolution of every species that has walked, dug, crawled, swum, flown or grown roots.

As we've already started to see, the structure of an individual's microbiome is partly determined by their genome and partly by the sum of our environmental exposures, known as the exposome. This becomes even more complicated because over time the microbiome will also influence the structure and function of the genome.

Describing the co-evolution of a complex superorganism made of ancient microbes required a new language. Lynn Margulis came to the rescue again, coining the term 'phylosymbiosis', which describes the co-evolutionary pattern between microbes and the humans they inhabit. In this holistic view of evolution, humans can be thought of as communities of interacting genomes made from multitudes of microbes, known as holobionts.

The holobiont concept states that microbial community members have contributed to our genetic, physical and social evolution within the rules of natural selection through both harmful and protective pathways. For example, pathogenic microbes employ common behaviours that promote their fitness at the expense of the host they inhabit, and the fittest of these hosts adapt to them. One of the world's most famous mummies is the 5,000-year-old murdered Tyrolean Iceman called Ötzi. His perfectly preserved body, dating back to the Copper Age, was found in the Tyrolean Alps and is still maintained today in a cold chamber at the South

Tyrol University. Ötzi's gut contained an ancient *Helicobacter pylori* genome, the bacterial symbiont responsible for causing gastric ulcers in modern man. It also contained pathogenic ancestor strains of *Clostridium perfringens* and *Pseudomonas veronii* that had the potential to cause illness.[16]

Non-pathogenic microbes have also evolved behaviours that manipulate host processes for their own survival. In some cases these have been so beneficial that they evolved with the host. Examples of this approach can be seen throughout nature: in return for a safe home, the clownfish cleans the tentacled anemone found in the Red Sea and the Pacific. It provides the anemone with nutrients and scares away other predatory fish, such as the butterflyfish. To do this, the clownfish has adapted to protect itself from deadly nematocysts, which are harpoon-like stingers on the anemone's tentacles used to capture prey.

The human gut offers many examples of similar symbiotic relationships that have co-evolved over similar timescales. *Bacteroides thetaiotaomicron* is much like a clownfish, living within the tentacle-like villi of the gut. It nourishes the gut and promotes its health and development, in return for a place to live. It's evolved to evade the host immune response by disguising itself in a classic piece of subterfuge: it camouflages itself by changing the type of immune molecules or antigens expressed on its cell wall, so that it can't be confused with a pathogen. Some of these relationships are so important to our survival that they are now encoded into our DNA.

Microbes with functions that regulate our physiology – such as digesting foods essential for survival, or the workings of an organ, or the development of our immune systems – are more likely to be passed from mother to baby, and therefore to be passed from human to human over generations (vertical gene transfer). This may occur in discrete social groups or geographically distinct regions where those microbes offer a specific survival benefit to that particular human population.

When man (and any other species, for that matter) has reached a fork in the evolutionary road, our bacterial partners have also skipped along with us through a process called co-speciation, if the phylosymbiotic partnership is strong. It is likely then that many of our ancient microbes were shared between Neanderthals and *Homo sapiens*. If a species of mammal becomes isolated by oceans or mountain ranges, co-speciation happens in isolation, because horizontal gene transfer between microbes of similar species of mammals becomes impossible and the microbiome lineages diverge. There are many environmental forces that influence the process of phylosymbiosis, but diet is one of the most important.

Feeding the evolutionary gut microbiome

Some 44,000 years ago, one of our ancestors crawled into a dark cave in the village of Leang Bulu' Sipong, on the southern side of Sulawesi, an island in Indonesia. In the flickering light of a torch, he or she intricately painted a pig and a small buffalo called an anoa in deep red and brown pigments. Today you can still clearly see that the anoa is being hunted with arrows, and the prehistoric hunters are holding rope to tie up the terrified beast. The implications of this exquisite work of art made by a prehistoric Banksy cannot be overstated. Its discovery in 2017 transformed the prevailing Eurocentric view of early man. For evolutionary microbiome scientists, this is also of particular interest as the microbiome of the inhabitants of Leang Bulu' Sipong would have been quite different from Europeans of the same epoch.

Our diets have, from the very beginning, shaped our co-evolution with microbial life forms. Human gut bacteria descended from ancient symbiotic bacteria that evolved from the great apes (hominids) into *Homo sapiens* over a period of about fifteen million years.[17] However, the image of the ancient carnivorous palaeo hunter consuming only freshly caught prey is inaccurate. Meat was, however, important for providing the energy needed for brain development, and it shortened our gut. To digest meat, our first carnivorous hominid relatives leveraged 150-million-year-old bacteria to help ferment it.

Protein has an important evolutionary role in regulating the gut microbiome because it contains a lot of nitrogen, which bacteria need for growth. Members of the phylum Bacteroidetes consume nitrogen in the large intestine more readily than other commensals do, and lowering dietary protein levels in mice reduces their faecal concentrations and vice versa.[18] This might explain why a high-protein Western diet has so fundamentally degraded our gut-microbial ecosystems. We simply consume way too much meat for our evolutionary microbiome to process.

However, the survival of early hominids (great apes) was also dependent on the metabolism of a very wide range of plant types found on the forest floor, and this was only possible because of herbivorous gut microbes that were 200 million years old. The modern omnivorous human gut is still made up of the relatives of these bacteria and their co-evolved metabolic functions, although we do not possess bacteria that are bespoke to our omnivorous preferences.[19] Cooking began somewhere between 1.8 million and 400,000 years ago, and this also dramatically shaped the evolving gut microbiome because it influenced pathogen and microbial consumption and nutrient availability.

The gut adapted to these changes, although our appendix is not (as Darwin proposed) an evolutionary remnant of our early plant consumption or something that has evolved with our dietary or social requirements. In fact it has been through at least thirty-two different iterations in its design across multiple species,[20] and it is an immunological organ in its own right. The appendix acts as a 'safe house' for commensal gut bacteria that train the immune system, which can prevent diseases by outcompeting dangerous pathogenic bacteria.

As *Homo sapiens* became dominant and began its multi-regional migration out of Africa, evolutionary changes in our oral and gut microbiome happened slowly because social structures were stable. Inevitably, gut bacteria began to diverge in their own evolutionary directions, depending on the geography and exposome. Metagenomic analysis of ancient faeces and bones lets us see how specific human microbiomes evolved during this time. Within dental plaques of hominids, for example, a core oral microbiome produced a mucus-like biofilm and – because it had an advantage for the host – it was maintained across primate species. And over time, species of oral streptococci developed the ability to bind to the human enzyme amylase, which breaks down starches, suggesting a close microbial co-adaptation with the host diet. By reconstructing the oral metagenomes of Neanderthals, scientists have identified some similarity with those of modern humans, although the diversity of the bacteria found in the mouths of our Neanderthal and Upper Palaeolithic ancestors has largely been lost. Some of the reasons for this are obvious and others far less so.

The evolutionary biologist Alan Cooper from the Australian Centre for Ancient DNA sequenced data from the dental plaques of Neanderthals from five regions of ancient Belgium. His team found that differences in diet were linked to an overall shift in the oral bacterial community, and suggested that meat consumption in particular contributed to substantial variation within Neanderthal microbiota. His team even found evidence for self-medication in fossils from an El Sidrón Neanderthal with a dental abscess and a chronic gastrointestinal pathogen (*Enterocytozoon bieneusi*).[21] That particular Neanderthal had been eating poplar trees, parts of which contain salicylic acid, a component of aspirin, and *Penicillium* mould. Genome comparisons of an oral archaeon called *Methanobrevibacter oralis* suggest that this microbe's modern lineage split from the Neanderthal some time after our last common ancestor lived, meaning that the archaebacterium was transmitted between different extinct human (hominin) species, possibly by kissing. Sex may be more than just the transmission of mammalian genes.

When *Homo sapiens* began farming 10,000 years ago, we started to ingest a lot more starch. Bugs in the mouths of farmers bloomed on the sugary films that lined their teeth as they started to chew cereal grains. As a result, *Streptococcus mutans* dominated, and dental caries soon followed. Fillings, fluorinated water and toothpaste have fundamentally changed the oral microbiome and the modern mouth now has a much lower diversity, which is good for our teeth and potentially less good for our gut. For example, some of the bacteria that are able to grow selectively in the mouth of urban individuals dining on a Western diet, such as the bacterium *Fusobacterium nucleatum*, are commonly found on cancers through the body, particularly in the colon.[22]

In our gut, an evolutionary superbloom occurred as our microbes gorged on the fibre and starch provided by farming plants, and the whole ecology of the gut changed again. Our reliance on cereals meant that the microbiome evolved enzymes that could help it consume this bounteous food source. Wheat consumption also shaped our immune systems, and today we are experiencing the breakdown of this ancient symbiosis. This is because modern industrialized farming practices have mutated ancient wheat plants cultivated for thousands of years, such as *Triticum monococcum* and *T. dicoccum*. As a result, these now contain greater quantities of a highly toxic form of gluten (a protein found in wheat) that possess a peptide called '33-mer gliadin'.

Coeliac disease is an inflammatory condition of the small bowel that causes abdominal pain, bloating, weight loss and malnutrition. It is caused by a hypersensitivity to the gliadin peptide, and it now affects more than 1 per cent of the world's population. The microbiome has influenced the evolution of this disease, because yeast such as *Rhodotorula mucilaginosa* and *R. aeria* co-evolved enzymes that break down and detoxify the immunogenic part of gluten, and some of these functions have been lost. Its most important role, however, may have been in the manipulation of the gut's immune system, because the microbiome influences both the body's ability to create antibodies to this protein and the severity of the immune response.

As society became increasingly urbanized, the *Homo sapiens* microbiome continued to deviate from that of its ancient relatives. The La Cueva de los Muertos Chiquitos archaeological site is located just north of Durango, Mexico. This eerie cave located in a dry desert valley was first excavated in the 1950s and contains infants and children, some of whom were thought to be sacrificial victims, and at least three adult burials. In deeper layers of this remarkable site, scientists have found ancient palaeofaeces known as 'coprolites', dating from the AD 700s to the early 900s.[23] Using metagenomic sequencing, they have been able to study how our microbiome has evolved in these samples over time. At phylum, family and species level, the compositions of the palaeofaeces samples were broadly similar to the gut microbiomes of present-day non-industrial individuals. This means that the basic structure of the gut microbiome had fully emerged. However, it was also possible to observe distinct differences between ancient and modern faecal microbiomes because the evolutionary process of co-evolution has not stopped.

Methanobrevibacter smithii, for example, is a dominant archaeon in the human gut, responsible for fermenting many of the fibres that we eat. It recycles hydrogen by binding it with carbon to produce methane and, yes, it is the farting bug (well, one of them). The scientists were able to establish that *M. smithii* began to diversify in the gut around 85,000 years ago, changing with our diet so that we gained the ability to metabolize certain foodstuffs – and the methane it produces also made our early settlements smell a little more fragrant. More worryingly, the research team also found that of the 181 most ancient genomes they studied, 39 per cent were from previously undescribed species, suggesting that some specific species within our microbiome have become extinct as humans have evolved. It is now possible that the microbiome extinction rate is climbing, and this is deeply worrying, as many of these strains of bacteria may have evolutionary conserved functions that protect us from chronic disease. This is because the loss of bacteria that co-speciate with us is more strongly associated with the development of immune diseases in humans – for instance, populations of bacteria of the genus *Subdoligranulum* are negatively correlated with inflammatory bowel disease.[24] Who is *missing* from our microbiome may be more important than who is there.

The similarity in the bacterial composition of gut microbiomes between people increases with geographical proximity and genetic relatedness. However, exposome-induced changes in the microbiome are dynamic, and evidence of our niche-specific co-evolution can still be seen today in indigenous populations that live in more isolated regions of the globe. For example, an analysis of the BaAka pygmy rainforest hunter-gatherers and their agriculturalist Bantu neighbours in the Central African Republic showed that, although their gut microbiomes were compositionally similar, they have evolved discrete functional differences. Hunter-gatherers, for example, have an abundance of Prevotellaceae, Spirochaetaceae and Clostridiaceae, with higher numbers of microbial virulence genes, and functions needed for amino-acid and vitamin metabolism. The Bantu gut microbiome, however, was more Westernized when compared to American gut microbiomes and was dominated by Firmicutes, which possess genes needed for carbohydrate and sugar metabolism.[25] Interestingly, the Bantu people also possess more microbial gene functions necessary for metabolizing industrial pollutants. Further studies of African populations in Tanzania and Botswana have identified that the microbiome of nomadic people that rely on domesticated animals is also phylogenetically discrete from that of hunter-gatherers.[26] Throughout the development of our species, our agricultural and survival strategies have therefore profoundly influenced the finely balanced process of co-speciation, and this in turn has had implications for our social and physiological evolution.

Some of the foods eaten by hominids may not have been that appetizing to you and me. Coprophagy or the consumption of faeces has been observed in captive and wild apes (chimpanzees, gorillas, orang-utans and gibbons). Primates do this because of its nutritional value and because it helps protect them against toxins in their diet. It is therefore possible that the first faecal transplants were much earlier than we think, although we have no definitive proof of this. This would not only have had potential health benefits in states of malnourishment, but could potentially have been a behaviour of evolutionary importance because it allows horizontal gene transfer between microbes. It is worth noting, however, that the great apes have a muted aversion to this gruesome practice, which could have also evolved to reduce their exposure to pathogens. The sensation of 'disgust' is a disease-avoidance behaviour that is found in many species and is of evolutionary importance in its own right.

Microbes and the art of war

Although no paintings of combat exist in the caves of Leang Bulu' Sipong, it is very possible that the men depicted in the hunting scene used the same weapons to defend themselves from other social groups that also coveted the tasty anoa.

Starvation and being preyed upon are powerful selective pressures that drive evolution in all animal species, and bacteria and microbes are no different in this regard. The gut is a dynamic and bustling ecosystem as busy and delicate as any Indonesian rainforest or ancient African plain – and thousands of species from the two other domains of life (prokaryotes and archaea) have been hustling for survival within us. *E. coli*, for example, has evolved over the last 120 million years into a cunning bacterium. It will play hide and seek, using its motor-propulsion system called flagella to swim away; or it will secrete a slime-like mucin, behaving much like squid do. It is continuously sharing genes, mutating and evolving, desperately and determinedly surviving despite the odds stacked against it. This fight is happening within you right now.

Most animals only engage in combat in small numbers, and most aren't particularly interested in killing or dying for mating privileges. But humans and bacteria are quite different. Our fights can involve many millions of individuals on each side, and these battles are often prolonged, intense and lethal. *The Art of War*, as written by microbes, makes Sun Tzu look like an amateur – they've developed chemical, mechanical and biological weapons of mass destruction. Sometimes they strike pre-emptively, at others defensively. They can choose to fight face-to-face, or to suppress their frenemies by exploiting or subverting the host's immune system. The bacterium *Pseudomonas aeruginosa*, for example, can punch holes through the walls of bacteria with its phage-like 'tailocins'. Others produce antibiotics, bacteriocins and other potent killing proteins; or they behave like silent assassins straight from a James Bond story, secreting lethal toxins into their sleeping targets using molecular hypodermic needles. Some microbes are able to engage in chemical warfare by making hydrogen cyanide.[27] In extreme circumstances, some bacteria will behave like suicide bombers and purposefully lyse or tear themselves open to release toxins that will kill their competition.

Other tools in bacteria's arsenal might have a survival benefit because they enable them to stay in position in nutrient-rich locations, for instance by sticking to membranes or by secreting slimy polymers, called biofilms, that smother their competitors. They even engage in information warfare through something known as quorum sensing. This is a molecular communication mechanism between bacteria that is triggered once a bacterial population reaches a certain density. Quorum-sensing chemicals work a bit like bacterial hormones, allowing colonies to control processes such as forming biofilm or to change their capacity to cause harm to their host. This is a global molecular language that bacteria use to communicate directly with their host. For example, some bacterial pathogens can use this network to sense human levels of the fight-or-flight hormone called adrenaline, which rises in response to an injury.

In other words, a human hormone determines if some bacteria produce toxins and ultimately how harmful they are to us. Adrenaline is not just stressful for humans.

The other major benefit of all this mortal combat is that it promotes biodiversity,[28] because efficient and effective strikes keep rival populations in check.[29] Whatever the evolutionary basis for this lethal combat, the interplay of mutation, recombination and positive selection of microbes created personalized and diverse niches within our bodies over a relatively short period of evolutionary time. But like a Sulawesi hunter, a microbe must also eat to survive.

The gut and the brain start a two-way conversation

The paintings in Leang Bulu' Sipong are extraordinary because they tell us that humans arrived in South-East Asia with the capacity for symbolic representation and storytelling. These magical murals also depict mythical figures, drawn as small humans with tails and snouts, suggesting that the artists may have had the ability to imagine supernatural beings. These two activities require a lot of brain power.

Many evolutionary theories have been put forward to explain the rapid growth of the primate brain, which doubled or even tripled in just three million years. Mechanical functions like walking, sensory functions like smell and sight, and talking demand a lot of neurones and brain mass. The size of primate groups, the frequency of their interactions with one another and the size of the frontal neocortex in the brain also correlate closely.[30] But these observations, and even spontaneous gene mutations, don't explain everything. It is also likely that the rapid growth in our cognitive capacity was influenced by a two-way conversation between the gut and the brain.

For a start, the rapid growth in brain size had to be fed: faster neuronal activity burns a lot of energy, and the brain uses about fifteen watts out of a typical body's total usage of seventy watts.[31] The brain couldn't have evolved at this rate without support from the gut microbiome, which regulates not only nutrient absorption, but also energy expenditure.

The hominid (great ape) predecessors of Leang Bulu' Sipong had a varied diet that not only sustained brain growth, but also ensured the biodiversity of the gut. We now know that the consequence of this is that the gut produces a vast catalogue of metabolites, which form the basis of chemical messengers known as neurotransmitters needed for higher brain function. In other words, deep mutualistic relationships between diet, bacteria and the gut shaped and supported functions within the neocortex that have enabled man to evolve beyond the great apes. Precisely how these variances change how our brain functions is still being defined, but as we will see in the following chapters, it does in part explain how our microbiome influences the way we feel, and it may even explain how we have evolved to think.

The future evolution of man

If diet were the only thing that mattered to the evolution of holobionts, then I, my microbes and planet Earth would not be having a midlife crisis. Subsistence strategies have of course had major influences on the evolution of our microbiomes,[32] but gaps remain in our understanding of the more subtle and rapidly changing environmental forces that have shaped the human microbiome. However, the man-made exposomes created with the birth of each new human civilization have exerted an increasingly destructive force on both our external and internal ecosystems. It's not only the planet that is warming up; it's our insides, too.

The greatest assault on microbial life forms has taken place in just the last seventy years, and as a result the human holobiont is profoundly unwell. These forces have been far greater than Ötzi's gut microbiome could have coped with; he may have endured a lot of comparative hardship, but Copper Age humans didn't have to deal with autoimmune disease and the diseases of progress. From this viewpoint, the popular gut-health palaeo diet is laughable. The Neolithic holobiont's gut would be so immune-naïve to the modern urban environment that Ötzi wouldn't make it through the day without contracting a lethal dose of man flu, and he would be as vulnerable to the obesity epidemic as the rest of us. Yet somehow the palaeo diet is worth about US $500 million, and the future evolution of the human holobiont stands at a crossroads.

We've co-evolved over many millions of years into a profoundly dependent relationship with our microbes that influences the function of all our organs and every aspect of our health. Our co-evolution is now speeding up, not slowing down, because this human–microbial partnership is deeply entwined with our environment. It's imperative that we understand how man-made changes in the planetary climate are influencing the terrestrial, marine and human microbiomes, and what a globalized human microbiome really means for our future health. This isn't a midlife crisis, it's an end-of-the-world crisis.

3. How Microbes Kill Us

How pale I look! – I should like, I think, to die of consumption . . . because then the women would all say, 'see that poor Byron – how interesting he looks in dying!'

Lord Byron

In 2011, anthropologists discovered a 5,000-year-old settlement in Hamin Mangha in north-eastern China. They excavated the foundations of twenty-nine houses, most of them simple one-room structures containing a hearth and doorway. In one of these houses, which measured just 19.5m², the scientists found the remains of ninety-seven bodies. The deceased were aged between nineteen and thirty-five; the cause of death is not clear.[1] The site had been abandoned and left for the gods or for anthropologists, whichever came first. Similarly gruesome discoveries from around the same time period have been made throughout Europe.

As Stone Age hunter-gatherers began farming, their lifestyles changed dramatically. Around 7,000 years ago Neolithic 'middle classes' moved into central Europe with their domesticated cattle and plants to settle, and they became comparatively sedentary. Families demanded faster food (which could at least be predictably harvested), entertainment and objects of desire. These came in the form of technological advances in pottery, animal traction, wheels and the discovery of metals. Trading became a necessity, and this brought an even greater innovation: money. Its first incarnation took the form of flint; this could be held in the hand or mouth, much like a one-dollar bill, which also has its own discrete microbiome.[2] The abundant communities of bacteria on paper money are sustained by contact with human skin, providing a record of human behaviour and health. Money and transport meant that trading networks could form and, for the first time in history, multiple independent human populations became connected. In turn, this meant that both pathogenic and symbiotic microbes could be shared between geographically dispersed populations of humans at unprecedented scale and speed.

Limited resources, competition, wealth and cohabitation led to politics and violence. The discovery of a 7,000-year-old mass grave in Schöneck-Kilianstädten, near Frankfurt in Germany, provides chilling evidence of this. Twenty-six bodies were found, and of these thirteen were children and ten were less than six years old at death.[3] They died violently, with evidence of blunt-force trauma and systematically broken legs, suggesting the first known instances of torture and corpse mutilation. We don't know who perpetrated the crime or why. We just know this was the start of a very gruesome practice, which only grew in popularity, and which continues to this day. Systematic death meant mortuaries became more sophisticated, and the necrobiome (the collection of microbes that decompose corpses) had more work to do. But violence alone doesn't explain the Neolithic decline or the events in Hamin Mangha.

Despite the combative reputation of this time, communal living became a reality and the property market 'boomed'. For the first time in history, urbanization began at a pace, and mega-settlements appeared in 6100–5400 BC in present-day Moldova, Romania and Ukraine. They were built by a population known as the Trypillia culture and could host between 10,000 and 20,000 people. But with scale came poorer sanitary conditions and the inevitable spread of pathogens. The Neolithic transition was accompanied by a rise in infectious diseases, driven by substantial changes in human ecology, geography, demography, housing conditions, hygiene and diet.

Modern endemic diseases such as tuberculosis were thought to have originated through zoonotic transmission from cattle about 6,000 years ago – although this may not be the case. Tuberculosis infections actually first appeared in early humans at least 35,000 years ago, and we may have been co-evolving with this bacterium for as much as 2.6 million years. This was long before it appeared in domesticated animals.[4] Its success in humans may have had less to do with our domestication of cattle and more to do with the rise in the size and density of Neolithic populations.

The late Neolithic and early Bronze Age communes were also the ideal breeding ground for a bacterium called *Yersinia pestis*, which causes plague. Our first record of early man's great nemesis has been found in the bones of Neolithic farmers in Sweden; a recent analysis of this lethal bacterium showed that multiple lineages of *Y. pestis* then mutated and expanded across Eurasia during the Neolithic decline. These most likely escaped through early trade networks, and it's possible that a pathogen such as this ended the lives of the families in Hamin Mangha. The ingenuity of Neolithic humans had taken them so far, but they hadn't accounted for bacteria. Rather than unhindered growth, their social innovations led to a sudden and unexpected net decline in the population.

Some pathogens have also exerted evolutionary influences on our risk of chronic disease. The most intriguing example is sickle cell disease (SCD), a group of blood disorders caused by mutations in the proteins that make haemoglobin. This leads to the formation of unusually shaped blood cells in the shape of a sickle, causing anaemia and extremely painful blocked blood vessels. Sickle cell disease is estimated to have originated more than 7,000 years ago and it was endemic in tropical regions of the globe and Africa.[5] The 'malaria hypothesis' states that this disease developed because it provided a survival advantage against malaria. This occurs because people who only inherit a single copy of the mutant haemoglobin gene from one of their parents and who do not develop symptoms of sickle cell disease (known as sickle trait) are resistant to the parasites *Plasmodium falciparum*, *P. vivax*, *P. ovale* and *P. malariae*, the parasitic pathogens transmitted by blood-sucking mosquitoes.

In the late Middle Ages, European settlements of all sizes were repeatedly struck by outbreaks of infectious diseases. Some of them, such as the Black Death in the fourteenth century, reached pandemic proportions. On this occasion, *Y. pestis* would go on to kill forty million people out of a total population of ninety million in

Europe. COVID-19 has been an indescribable tragedy, but the Black Death killed 45 per cent of the entire European population. And the variola virus that causes smallpox has been responsible for the death of 300 million people. It's thought to have first emerged about 1,700 years ago, during the turbulent period of the fall of the Western Roman Empire, which saw people migrating across Eurasia in unprecedented numbers.

Humans have always carried pathogens. The success of every civilization – from the Aztecs, Chinese dynasties, Incas, Greeks and Mayans to the Mongols, Persians and Romans – has been influenced by pathogenic microbes that bloomed and receded in parallel with their geographical and social influence. The nineteenth and twentieth centuries were no different. In the twenty-first century, however, pathogens like SARS-CoV-2 can rip through a global, connected society with unprecedented speed.

True crime

Approximately sixty-nine million people across the globe died in 2021, a rise from 57.94 million in 2019. Much of this can be explained by the COVID-19 pandemic, although in any given year, outside of a global health crisis such as this, about one-third of deaths are caused by communicable infectious diseases that are passed between humans. Of the millions of species of microscopic life forms that exist on this planet, only about 1,400 are known to be pathogens, and of these, 500 are capable of human transmission. Just 400 cause serious harm to humans (the most effective killers are tuberculosis, HIV, malaria, measles and influenza), and fewer than 150 have the potential to cause epidemic or endemic disease.[6] Your risk of contracting these diseases varies greatly, depending on whether you live in the developing or developed world and, thanks to modern medicine and the widespread use of vaccines, the majority of these pathogens are now far more likely to maim or torment you than kill you.

In the nineteenth century, however, there was a much greater statistical chance that a microbe would be the cause of your death.

The top three causes of mortality were pneumonia, tuberculosis and gastrointestinal infections. Pneumonia kills through a combination of suffocation and sepsis. Chest pain and shortness of breath are caused by stiff, inflamed lungs, before pus collects in the tissue of the lung, preventing oxygen from being absorbed into the blood – leading to suffocation. Sepsis describes a complicated and deadly process by which bacteria escape from a confined space in the body, such as the lung, into the blood supply. As they replicate, microbes hack into and overwhelm the immune system, ultimately uncoupling it from its feedback control mechanisms. With the safety-switch off, our defences run wild and the body begins to attack itself; the blood stops clotting, and the body's organs begin to shut down one by one. It's the dreaded and common pathway for many different types of pathogenic microbes. Even microbiota commonly found on our body, like *Staphylococcus aureus*, can do this to us. Today, sepsis kills one in five people, and it disproportionately affects vulnerable populations such as newborns, pregnant women and those living in poverty.

Pneumonia was a prolific and violent killer, but public health strategies to combat it were slow to materialize in the twentieth century: its cause was simply not known. It's a similar story with *Mycobacterium tuberculosis*, which also fills the lungs with pus, although it kills through a slower process of abscesses, ulcers, blood loss, weight loss and exhaustion. Hence 'consumption', as it was typically referred to during the eighteenth and nineteenth centuries. By the 1800s there was an epidemic of TB in Europe; more than four million deaths from the disease were recorded in England and Wales between 1851 and 1910.[7] With over one-third of the dead aged between fifteen and thirty-four, TB also gained the moniker 'the robber of youth'. Other names for the condition included 'the white death' – a reference to the anaemic pallor of its victims – and, my personal favourite, 'the King's Evill', so named because it was believed TB could be cured by the King's touch. Things haven't changed all that much in the interim. During the COVID-19 pandemic, many people sought the digital touch of modern royalty – social-media influencers or celebrities. If the American podcaster and martial artist Joe Rogan says that he won't take the COVID-19 vaccine, then millions follow the king.

Gastrointestinal diseases complete the death trio of the nineteenth century. Cholera became a global killer, with frequent large-scale epidemics in European cities. In 1854, an outbreak of cholera in Soho, London, killed 500 people. Through meticulous observations, Dr John Snow – now considered the father of modern epidemiology – identified contaminated water from the Broad Street pump as the source of the disease. The cure? Simply remove the handle. (If you happen to find yourself in London, you can still visit the site today.) In 2020, diseases that cause diarrhoea were still the eighth most common cause of death, with norovirus the biggest killer among them, particularly in children and those living in Asia and sub-Saharan Africa.[8] Gastrointestinal diseases are as preventable today as they were for John Snow, but we can't seem to find the pump handle. Politicians in the 1800s were unforgivably slow to act in the health crises they faced, and today not much has changed.

The microbiology renaissance

Despite the importance of microbial pathogens to our survival, almost nothing was understood about the perpetrators of these crimes until the nineteenth century. Through the miasma, fear and grime of the Industrial Revolution, some of the greatest medical and scientific minds began trying to solve the problem of infectious disease. Today, we take for granted that pathogens – microbes with the potential to cause disease – exist. In the early 1800s, the idea that a microscopic life form could be responsible for so much human suffering was inconceivable. There was, however, some speculation that cholera and the Black Death were driven by microscopic beasts, and the 'spontaneous generation' theory – which suggested that, through divine intervention, living creatures could come from non-living matter – became popular. For many believers and religious groups to question this theory was heresy.

Louis Pasteur, perhaps the greatest microbiologist of all time, was, like many geniuses, distinctly average in his early life. He initially struggled with chemistry, a subject that he would later excel in, and failed to gain entry to the elite École normale supérieure in Paris on first applying. He began to address the topic of germ theory early on in his career – and was virtually ignored for the first twenty years of his professional life. But, like all greats, he possessed two key qualities: persistence and an open mind.

Pasteur understood that if he could describe the function of microscopic life forms, whether they were fermenting yeasts or fungi-killing silkworms, he could meaningfully save not only whole industries, but also lives. He disproved the theory of spontaneous generation through a series of technically brilliant experiments which showed that the 'vital principle' existed in air, and that this was responsible for the 'spontaneous' appearance of microbial life in sterilized broths – they'd hitched a lift on dust particles that travelled in the laboratory air. His work proved that life can only arise from pre-existing life, and with this he dealt a lethal blow to ancient dogma.

His breakthrough was followed by an explosion of discoveries that began to define how microscopic life forms cause disease. The modern age of preventative medicine had begun, and with it came the true birth of vaccination.

The four postulates of the apocalypse

Thanks to the strides taken in microbiology throughout the nineteenth and twentieth centuries, today we live in a world where, outside pandemic years, the biggest killers of adults are no longer infectious diseases. Deaths from non-infectious diseases take forty-one million people each year – more than seven out of ten deaths worldwide. Proportionally, as many people die each year in the United States from ischaemic heart disease as they did from pneumonia in the 1900s,[9] and it causes the death of 16 per cent of the world's population. The second-biggest killer globally is stroke (11 per cent of global deaths), then chronic obstructive pulmonary disease (6 per cent).[10] As well as killing, these diseases cause devastating disability and economic disadvantage. We also have a new burden of mental-health disorders, disorders of the immune system and chronic diseases of old age such as dementia, which were of minor importance 150 years ago as people simply didn't live that long.

Until recently, microbes escaped the blame of the medical profession in causing non-infectious diseases – the very definition of a pathogenic microbe rendered the link impossible. The reasons for this date back to 1882, when the great bearded scientist Robert Koch published what would become a definitive paper on his discovery of the bacterium responsible for the 'King's Evill' – namely, *Mycobacterium tuberculosis*.

The paper, which made Koch an instant international hero, celebrity and, ultimately, Nobel Laureate, formulated the framework that continues to inform our thinking on how microbes cause disease. (In truth, many other leading scientists were describing similar frameworks, but Robert Koch's name seemed to stick.) His four postulates state that for a pathogen to be causally associated with a disease:

1. The same organism must be present in every case of the disease.
2. The organism must be isolated from the diseased host and grown in pure culture.
3. The isolate must cause the disease when inoculated into a healthy, susceptible animal.
4. The organism must be re-isolated from the inoculated, diseased animal.

Even Koch knew that these were not watertight rules and that exceptions would inevitably exist. And he certainly never intended for his postulates to be used when considering chronic disease. And yet, for many years, his postulates were the standard to which microbiome scientists were held. As a result, proving the role of the microbiome in causing chronic diseases was challenging at best, and conceptually impossible at worst.

Barry's marvellous medicine

In the 1980s, two Australian researchers – Barry Marshall and Robin Warren – became the first scientists to describe a new curved bacterium, *Helicobacter pylori*, in the stomach of patients with gastric ulcers.[11] They argued that gastric ulceration wasn't caused by acidity in the stomach, as medical dogma dictated, but by a microbe. It was a seismic discovery that challenged the whole basis of chronic disease.

No one believed them, and so in 1985 Barry Marshall took drastic action. He persuaded his colleagues to perform a gastroscopy (whereby a flexible camera is passed through the mouth into the stomach) and to take biopsies from his stomach to prove that he did not have the bacteria present. Then he drank a live culture of *H. pylori* to prove that it caused acute gastric illness. He was right, because *H. pylori* causes more than 90 per cent of duodenal ulcers and up to 80 per cent of gastric ulcers, and it's also associated with a raised gastric-cancer risk. Today, almost all ulcers in the stomach are not treated with surgery but with antibiotics and antacids. However, this hasn't been without controversy, and some people are now starting to ask if *H. pylori* may have been unfairly stigmatized as a pathogen, because it also possesses some beneficial roles for the maintenance of gut health.

Some microbes are intent on causing you harm and allow no room to negotiate. The Ebola virus, for example, is going to do its very best to kill you – and it will be an unpleasant and bloody demise. But other microbes might be 'friendly' until placed under conditions of extreme stress, when they're forced to resort to self-defence. Others work together in teams across vast networks and exert their pathological effect in a far-away organ such as the brain, using discrete chemicals that might be produced by many different species of bacteria. Finally, in some cases it is the absence of an essential microbe, such as bacteria of the genus *Subdoligranulum*, that causes harm, and if you don't know what you're looking for, this can be incredibly hard to prove.

Modern microbiome science has developed a new lexicon to talk about how bacteria and microbes and humans interact to promote health and cause disease. A *symbiont* is one organism living closely with another, and this relationship is usually – but not always – peaceful. In *mutualistic* relationships, both the microbe and the host benefit from the relationship. Microbes within our gut, for example, share fuel and critical resources through a process known as 'cross-feeding', and this represents a cascade of effects across a network. In *commensalistic* relationships, one partner benefits while the other does not. This is a bit like being in a relationship where only one of the partners is satisfied; if one person is more unhappy than the other, then it is not going to be harmonious. *Parasitic* symbionts gain a benefit while harming the host; there's no better example of this than the human papilloma virus. *Amensalistic* symbionts, however, are the most selfish of all: they harm us without deriving any benefit from the relationship. These sadists are destructive, with no apparent purpose.

It isn't always easy to work out whether an organism is mutualistic, commensalistic, parasitic or amensalistic. *H. pylori* is a commensal symbiont that we've co-evolved with, to maintain the health of the foregut (even Ötzi had this bacterium in his gut), and because it exists in a complex mutualistic network, attempts to eradicate it with antibiotics have had unintended consequences – for example, the rise in new chronic diseases associated with unnecessary antibiotic use, such as asthma, allergy[12] and even coeliac disease.[13] What's more, this effect may be multi-generational, as the presence of maternal *H. pylori* protects against asthma in the offspring.[14] In my view, *H. pylori* is most certainly not a ruthless pathogen – and we shouldn't be blindly eradicating it without understanding its wider role in our health or the consequences of antibiotic use for its treatment.

What we do know for certain is that the discovery of *H. pylori* ensured scientists all over the world started to ask whether microbes might be responsible for causing chronic diseases like cancer and dementia. A new chapter in microbiome research had begun.

The microbiome postulates

Nineteenth-century microbiological doctrine is no longer fit for twenty-first-century purposes. We need a new standard for defining the importance of microbes to our health and non-communicable disease – one that moves beyond Koch's postulates. We also need a structure that helps us understand the mechanisms through which the microbiome may change the effect of a medicine, a behaviour or an experience. I humbly submit the microbiome postulates for your consideration:

1. *The same microbe or network of microbiota and their molecular functions (a microbiome!) must be present in an individual or in an anatomical niche before a disease begins, if it is to be causally associated with that predefined state of health or disease.*

    This sounds obvious, but our microbiomes are unique to the individual, to their anatomical location or organ and they even change as a disease progresses. The importance of this postulate can be illustrated through the example of bowel cancer. Cancers within the bowel have what have been termed 'passenger' microbes that exist in or on the tumour, and these are detectable at the time of diagnosis. This doesn't necessarily mean they're the cause of the cancer, which may have happened many years before the diagnosis and sampling. So as well as 'passenger' organisms, there are 'driver' organisms, which might have originated as part of a very different ecosystem, under different environmental conditions, when the colon would have appeared healthy to the naked eye. An example of a passenger organism is *Fusobacterium nucleatum*, a bacterium typically found in the mouth, but which has now been identified in gastric, colorectal, pancreatic, liver and even breast cancers.[15] The sexually transmitted virus HPV is, however, a well-established driver of cervical cancer, and it has a dynamic relationship with the passenger microbes that it lives with in the vagina. The real question is: how far do we have to go back before the start of the disease to identify the culprit or culprits? That question isn't easy to answer – we might have to go back a whole generation. For instance, there is evidence that maternal obesity in pregnancy increases the risk of bowel cancer. So we need a second postulate.

2. *The microbiome and its functions must be measurable at multiple time points in the same individual or group of people to demonstrate causation or treatment response.*

This postulate acknowledges that a chronic disease or its treatment will also alter the microbiome. As a cancerous tumour evolves and mutates, the physiological condition of the tissue also changes, so the cancer passenger microbiome changes too. The levels of tissue oxygenation or blood supply will also vary as the tumour grows, and this creates a niche for microbes that prefer low-oxygenation states. Some microbes might also have a survival advantage by living within the cancer, and in turn this oncological symbiosis helps the cancer meet the high energy requirements of immortal cancer cells. A disease like cancer doesn't have a static microbial community but a dynamic and highly individualized ecosystem, and this is also important if you're trying to treat it with a drug. To further complicate matters, cancer has a discrete zone around it, known as the 'tumour microenvironment', which hosts immune cells and other important housekeeping cells that influence and support cancer growth. The microbiome also regulates cancer growth by interacting with the immune cells of the tumour microenvironment, and it's possible that these microbes are more representative of driver organisms.

Because there's such great variability at a molecular level in the types of cancer that exist, and in the microbiome, this postulate also allows an individual to serve as their own control, and for microbiome studies to account for shifts in microbial abundances caused by many confounding variables, such as diet, lifestyle, travel or unrelated illness.

3. *Microbiome experiments in the laboratory (known as in vitro analysis) must replicate the entire microbiome and its functions over representative time courses to demonstrate causation.*

    Animal models of human disease aren't just ethically and morally dubious – they don't work particularly well in the analysis of the human microbiome. That is because it is very hard to stably reconstitute a microbiome's structure and function from one species in another. It's not possible for a laboratory animal to truly experience a human environment; you can take a rabbit to KFC, but it won't eat a family bucket of wings and wash it down with two litres of Coke and a cigarette. As a result, very few animal experiments on the microbiome have successfully translated into humans. Wherever possible, microbiome studies of human disease should be performed in humans. This is challenging – humans don't like being poked or prodded for the sake of science. What's more, the simple act of accessing the gut (via a colonoscopy, for instance) changes the conditions of the gut by introducing toxic oxygen or by removing faeces, both of which change the microbiome. This problem might be solved in the near future with the development of synthetic models that allow human stem cells to be grown in a sophisticated three-dimensional cell culture known as an 'organ on a chip'; this culture contains all of the anatomical and physiological functions needed to sustain a microbiome. Bioinformaticians are also creating computational models that will give us much greater insight into how whole microbiomes function and interact.

4. *If an obscure claim about microbiome research only exists on social media or it was made by a 'health influencer', you*

probably should ignore it. Don't fall for people peddling quick answers and easy cures.

The microbiome is a complex system. The job of a scientist is to model and handle the uncertainty of these systems and to generate repeatable observations for well-defined hypotheses that don't answer all of the questions, but only the question that the experiment was designed to answer. We're on a journey that requires patience and Pasteur's open mind. But the digital world wants answers now and it wants certainty. If it doesn't get this, it fills the spaces in scientific reasoning with hype and conspiracy theories. The microbiome is going to give us new and important therapies that will transform human health, but this will take time.

The Anthropocene decline

Over the last 200 years or so, the importance of the microbiome to human health has been obscured by the breakthroughs of early microbiology. We're living longer because we've had to deal with relatively fewer pathogens, but we're not living happier, because the microbes we're left with don't keep us well. We can't rely on outdated microbiological doctrine to explain modern pandemics – of either pathogens or chronic diseases.

In 1882, Koch had no way of knowing the lung, gut or skin microbiome existed, and he couldn't have known that the interaction between bacteria, viruses, phage, fungi, protozoa and helminths in one organ defines, in part, the immune response to an acute or chronic infection with a pathogen in another. The metabolic functions of gut bacteria, for example, are now strongly implicated as a cause of atherosclerosis – the process of the furring of our arteries that causes strokes and heart attacks.[16–18] The scale of the chemical superhighway that originates in the gut can't be underestimated; the gut microbiome creates hundreds of thousands of different types of metabolites and proteins that modify disease risk. In a study of UK twins, the microbiome was found to be involved in the production of over 70 per cent of faecal and 15 per cent of blood metabolites.[19] It's the production of these small molecules by networks of bacteria that defines the risk of disease in distant organs.

Human–microbiome relationships are dynamic, and the microbiome exerts its influence on our health through a network of microorganisms and a lifetime of gene–environment interactions, some of which didn't exist in the 1800s. In the intervening century, we've finally started to change our understanding of what constitutes a pathogen.

There are striking similarities between the Neolithic decline and the unofficial epoch that we are currently living through, which is named the *Anthropocene*. The same major socio-political, economic and biological forces that caused society to be less resistant to pathogens 10,000 years ago are happening again, but on a previously unimaginable scale. We're even more sedentary, we covet wealth, we live in massive cities, we're connected beyond our wildest dreams. We're less likely to maintain the diversity of our internal ecology and we've become obsessively hygienic and isolated. This time, however, there is a much greater danger: the Anthropocene is so named because it acknowledges the impact that humans have had on our planet's climate and ecosystems. Global microbial ecosystems are collapsing around us and within us.

Today, the greatest risk to urban dwellers in most developed countries doesn't come from pathogens, but from absent or misdirected symbionts and amensalists causing pandemics of chronic diseases like autoimmune disease, obesity, allergies, asthma and mental-health problems. Most of us don't die young, drowning in our own pus, but we die in hospital when our minds have failed and our expensive drugs have stopped working. Bacteria are no less responsible for our health, happiness and death than they were 5,000 years ago. It's just that our perception of their importance has changed.

Germ theory won't save us, but microbiome theory might.

4. Inflammation

In the small hours of 13 August 1961 the German Democratic Republic began to seal off the eastern sector of the city of Berlin. Almost overnight, a concrete wall covered in barbed wire spanning twenty-seven miles was constructed. It was built to prevent East German citizens fleeing from a totalitarian state. Life in the East was bad enough that between 1961 and 1989 more than 5,000 made successful escapes over and under the wall, and 140 people were murdered trying. Almost 40 per cent of the population in East Germany worked in industry and factories. They had limited access to fresh fruit and vegetables, and many families relied on single-room heating with fossil fuels. There were also major social differences between citizens of the GDR and their Western counterparts. Contraception was normalized, women worked, they donated breast milk to milk banks, and single-child families were common. Some East Germans drank Vita Cola and watched westerns, while the West Germans had access to Hollywood blockbusters, Coca-Cola, fast food and could travel where they pleased. Whatever East Berliners did, someone was always watching.

As suddenly as the wall went up, it came down. On the night of 9 November 1989, in response to demonstrations, an East Berlin party official named Günter Schabowski announced what were supposed to be limited travel reforms to stunned reporters. However, he bungled the press release, and the orders were misinterpreted as permission to open the gates. The guards on the wall were overrun by ecstatic crowds. Within a year of the fall of the wall, German reunification was complete.

Epidemiologists watched the reunification process with great interest: the controlled cultural cross-pollination represented a unique opportunity to study the impact of environmental factors as a cause of Western diseases. East Berliners suffered from the diseases of deprivation and malnutrition; post-unification, there was a rapid improvement in life expectancy (four years for females and 5.7 years for males[1]). But East Berliners also had much lower rates of immune-mediated diseases, such as hay fever and pollen allergies. Within just two decades of reunification, East Berliners still suffered from comparatively higher rates of poverty, but they had gained the same risk for these allergies as West Berliners.[2] These important observations reflect what is happening across the planet today: an explosion of allergic rhinitis, or hay fever, and immunological disease in response to the relentless globalization of Western culture.

Allergy is now the most common chronic disease in Europe, and its rising prevalence is startling. By 2025 half of the entire EU population will be affected; seven times as many people were admitted to hospital with severe allergic reactions in Europe in 2015 as in 2005. Around 300 million people suffer from asthma, and 250,000 will die prematurely from this condition each year.[3] Food allergy cases have also risen by as much as 50 per cent in the past decade, with a 700 per cent rise in hospitalizations due to anaphylaxis.[4] The consequences for young people of this dramatic change are significant – and the change isn't solely down to better diagnosis. We urgently need an explanation for the rapid rise in the constellation of chronic diseases that disproportionately affect young people.

War games

As a child, one of my favourite films was *WarGames*. It was released in 1983 and starred the ageless Matthew Broderick as a high-school hacker who inadvertently breaks into a military artificial intelligence (AI) supercomputer known as the WOPR (War Operation Plan Response). Broderick and his girlfriend, played by Ally Sheedy, discover a list of what appear to be computer games. Like any teenager would, Broderick decides to play a game of 'Global Thermonuclear War', and he chooses to simulate a Russian attack on mainland North America, starting with Las Vegas. The 1980s computer graphics seemed so sophisticated at the time, and back then AI could only be imagined by Hollywood. What Broderick and Sheedy don't realize is that this is actually a piece of top-secret software belonging to the US government, which interprets their gaming as a real-world nuclear attack. Broderick is arrested, and although he convinces the military that it is not a real Russian attack, the WOPR computer, made from very convincing red-and-white flashing LEDs, doesn't believe him. Speaking in a monotone humanoid voice, it calmly overrides the feeble humans and prepares to launch a real counter-strike at Russia. Lots of large monitors in the military bunker project the end of the world. A panicking Broderick escapes, runs across America and tracks down WOPR's mad scientist creator, who is living as a bitter and cynical recluse. He brings the AI boffin back to the military's war room, and the computer scientist saves the day by challenging WOPR to a game of 'tic-tac-toe' (noughts and crosses to me) because it teaches the intelligent computer the concept of no-win scenarios. WOPR ultimately concludes that Global Thermonuclear War is 'a strange game' and that the only winning move is not to play. War is averted and everyone involved learns that AI is a dangerous thing, and nuclear war is bad. We can only hope that *WarGames* is required viewing for all dictators with itchy fingers and a nuclear button.

The film is a good analogy for autoimmune disease – our immune system is the WOPR supercomputer. This highly complex, semi-autonomous system was originally built for defence against pathogens, but now it has been hacked and inadvertently put on a path of mutually assured destruction. Your microbial pathogens and your symbionts are two global superpowers that are fully armed, with their molecular weapons pointed at each other, and you and I are better-looking versions of Matthew Broderick and Ally Sheedy; we don't really understand what is happening, but we are desperately trying to avoid a global thermonuclear war of chronic disease that we may, or may not, have accidentally started.

In its most basic form, the immune system consists of two intricate and intelligent parallel networks: our innate system is primed to respond to any intruder without any prior knowledge, while our adaptive system (which evolved over millions of years with help from viruses) is able to learn friend from foe. The gut is the central command bunker of the combined immune intelligence agencies, and it houses the largest collection of immune cells, lymph nodes and lymphoid tissues in the body – and, with it, the largest microbiome.

Our injured or dying cells release proteins and molecules that independently activate the innate immune system to promote healing. The innate immune system is clever because it can distinguish these molecules from the lipopolysaccharides, sugars, proteins and metabolites made by microbes. The cells of the innate immune system line our gut, and they can detect these molecules using pattern-recognition receptors. Once activated, these receptors trigger a master switch within immune cells, known as the inflammasome. This starts up the inflammatory response within the cell, which in turn releases a vast cascade of soluble molecules that swarm around the body, ringing the alarm, calling other immune cells into battle. While the immune system does its job by removing the intruder and repairing the damage, we feel unwell and we get a temperature.

Unlike the innate immune system, which serves as a defensive reflex, the adaptive immune system is a precision-strike unit. It may be slow, but it has astonishing powers of memory, and it never forgets an antigen. If the adaptive immune system were a smartphone, it would have tracked my every movement without my consent, and it would know the name of every symbiont, pathogen and amensalist that has ever visited my gut. But before the lymphocytes of the WOPR adaptive immune system can launch an immunological strike, they must first be educated and primed. Like any good intelligence agencies, such as Britain's MI5 and MI6, the innate and adaptive immunological agencies share information about local and distant threats. Particular types of cells of the innate immune system called macrophages, for example, continuously eat up and ingest antigens in the environment and, on bended knee, present their findings to their immune masters, called white blood cells (B and T lymphocytes).

When activated, B cells may produce up to five major classes of antibodies, which are Y-shaped proteins that can bind to antigens present on pathogens. These either signal to the cells of the immune system so that they can ingest and destroy an intruder, or they can directly kill themselves. Immunoglobulin A (IgA) lives in our saliva and mucus and targets pathogens for destruction, and it is a first line of defence. Our most ancient antibody, called IgM, rises quickly in the blood when we experience a new infection, to suppress the intruder. After time, the adaptive immune system develops more targeted IgG antibodies that take over this role. Your body then remembers to make the same IgG antibodies in case you meet the same pathogen ever again, and that way it doesn't have to go to the trouble of making IgM. These are the antibodies that are made when you develop immunity after having a vaccine. IgD is a signalling antibody that helps produce IgM. IgE binds to allergens and triggers specialized cells to release chemical transmitters such as histamine, the chemical that causes itching and swelling of our tissues when we have an allergic reaction.

Our ability to artificially synthesize subclasses of these immunoglobulins as medicines has been one of the greatest single advances in modern medicine in the last two decades. 'Immunotherapy' means that we no longer need to make our own immunoglobulins; we can just prescribe them. These blockbuster drugs have transformed the treatment of chronic diseases of the immune system, such as rheumatoid arthritis and inflammatory bowel disease, and they are transforming outcomes in many different cancers. As we shall see, the deep evolutionary relationship between the adaptive immune system and our symbionts also explains why the microbiome holds the key to the success and failure of these new medicines.

Collectively, this whole innate and adaptive network functions as a continuous surveillance system for the presence of microorganisms in tissues. The microbiome establishes a form of immune 'tone' that keeps this whole system in balance. However, over just a few hundred years, the innate immune system has found it increasingly hard to distinguish our body from symbionts, pathogens and environmental toxins, and this immune tone has been lost. The result has been friendly fire, chronic inflammation and an epidemic of new diseases.

Inflammation doesn't describe a stable, homogeneous state – it's a dynamic event that progresses through different phases of senescence and volatility. Many of its infinite pathways overlap or are triggered simultaneously by pathogens, toxins and allergens. If it is left switched on for too long, or if inflammation escalates in an unregulated way, it ultimately begins to damage our tissues, joints and organs, causing disease. Sometimes this damage occurs over weeks, months or years and at other times it happens in seconds – for example, in the case of anaphylactic shock, when an allergen causes a collapse of the vascular system, the narrowing of airways and death, if not treated with adrenaline. Because of this, the immune system has also developed a series of checks and balances that are designed to keep it under control, in the form of anti-inflammatory molecules that suppress inflammation through similar networks.

Microbiome science has been so revolutionary because it has given us a completely new perspective from which we can study these inflammatory and anti-inflammatory interactions. It has provided deep insight into how our symbionts have contributed to the design of these control systems and their ability to hack into the immune WOPR computer through the back door (just like the impish Matthew Broderick). They can access every part of the broader immunological network, and if ambivalent amensalist microbes (like HPV) are given a chance, they can choose to play global thermonuclear war at any time and cause a cancer. Symbionts can equally soothe the immune system and deactivate it. The gut is so dependent on its symbionts to maintain the inflammatory status quo that, if it is left without them even for a short time, it cannot keep the immune system in the 'off' position. For instance, after very large doses of antibiotics, the gut becomes angry and raw because of this phenomenon.

The close evolutionary relationship between our microbiome and the cells of our immune system may also explain why FMT appears to be so effective in reducing the immunological complications of bone-marrow transplants used in the treatment of cancers of the blood such as leukaemia. Clinical trials of FMT that improve the diversity of the gut appear to reduce the chances of the immune system inadvertently attacking itself (this phenomenon is called 'graft vs host disease') and it improves the likelihood of this life-saving therapy working.

Trying to understand why this happens is complicated, because the gut's immune system has developed evolutionary relationships with hundreds of thousands of species from all three domains of life and with viruses, and these relationships intersect at multiple points across the innate and adaptive immune systems. Phage and viruses of the gut's 'dark matter' are likely to be significantly under-rated players in maintaining the immune system, but at the moment the relationship is almost completely undiscovered. It is also worth remembering that very few theories that attempt to explain the rising tide of immune disease place much significance on the more minor players in the microbiome. Just because they are not very abundant, that doesn't mean they are not important.

Nematodes like roundworms and threadworms, for example, don't generally kill us, but our immune systems can't eradicate them; we've had no choice but to learn to put up with them, and their presence in the human body might be asymptomatic or it might be debilitating. In patients with blood nematode infections, the inflammatory response is down-regulated to avoid excessive tissue damage. As a result, worms have been used as a successful therapy for inflammatory bowel disease. Multiple sclerosis (MS) is a devastating degenerative condition that can affect the brain and spinal cord, causing a wide range of potential symptoms, disrupting vision, arm or leg movements, sensation or balance. Clinical trials of helminths in MS patients have demonstrated that they are able to slow the progression of this disease, and that their administration causes a rise in the levels of a particular class of T cells (known as regulatory T cells) that recognize nerve cells and suppress the adaptive immune response that damages them.[5] Hookworm may yet serve as a therapy for this debilitating and cruel condition.[6]

There may also be important implications for the interaction of microbial species and our immune system in its response to acute illness caused by pathogens like viruses. Evidence from African researchers indicates that significantly fewer COVID-19 patients suffer from serious symptoms than in the industrialized world. A major hypothesis suggests that this can be ascribed to intestinal parasites, which mute the hyperinflammation associated with severe COVID-19. In one Ethiopian study of 751 patients, almost 38 per cent were co-infected with one or more parasites, and this cohort was statistically far less likely to have severe COVID, regardless of their age and comorbidity.[7] What's more, this observation was consistent with any parasite, including protozoa or helminths. Despite similar observations from other international cohorts, in 2021 a study was published in a pre-print form (meaning it was not peer-reviewed) by a team from Benha University in Egypt, suggesting that an anti-parasitic medication called Ivermectin was useful in the treatment of COVID-19.[8] These dubious data were seized upon by anti-vaxxers and Facebook user groups and the medication was pushed across social media as a cure. This pre-print study was ultimately withdrawn, but even if the Benha data were credible, an anti-parasitic drug would make no mechanistic sense, given that it's the parasites that appear to be limiting the severity of a deadly viral pathogen.

Microbiome–immune interactions are the key to unlocking some of the biggest threats to human existence, from infertility, cardiovascular disease, obesity, cancer, infectious disease and even our mental health. The problem is that the body's immune WOPR computer is not distractible with tic-tac-toe, and we haven't negotiated effectively with the symbiotic and pathogenic microbial superpowers that are armed to the teeth and looking for a fight.

The question many scientists are now starting to ask is whether our microbes also explain why we have a rising tide of allergies.

Extinction rebellion

In 1989, Professor David Strachan at St George's Hospital in London observed that children with older siblings were less likely to get hay fever and eczema. He noted that 'declining family size, improvements in household amenities, and higher standards of personal cleanliness have reduced the opportunity for cross-infection in young families'.[9] This was an exceptionally important observation that suggested a role for microbes in this chronic disease, but it didn't explain why some children in poorer communities might still have asthma, or why East Germans had higher rates of atopic eczema than West Germans before reunification occurred. It couldn't only be about improving standards of hygiene. The new science of the microbiome, however, is starting to uncover the nuance in this proposed mechanism.

More contemporary ideas on this theme, such as the 'old friends' or 'biodiversity' hypotheses, argue that because we have so closely co-evolved with organisms, we have actually come to rely on them to educate our immune system.[10] If we experience any disruption to the ecology of the gut microbiome, we become unable to distinguish friend from foe, and the WOPR computer begins its game of thermonuclear war when we meet a potential allergen. It's not just that a lush and diverse gut ecosystem is important for healthy immune development, it is more that these exposures need to happen in early childhood so that the adaptive immune system can develop appropriately.[11] In fact they may have to happen even earlier than this.

In 2009, Martin Blaser and Stanley Falkow presented their theory of the disappearing gut microbiome.[12] They argued that the absence of ancestral indigenous microbiota over many generations explained the rising prevalence of chronic diseases and allergies. Because the damage caused by the environment to our internal ecology is embedded in the maternal microbiome, it is passed from mother to child – and the impact of these environmental influences are therefore cumulative. In other words, the core maternal microbiome and its long list of critical functions that we need for good health become diluted with each generation. Blaser and Falkow hypothesized that the health of the parental microbiome and the method by which we bestow this on our children is equally critical to the parental stewardship of their children's microbiome during the first few years of life. This is why so many scientists are now interested in the long-term consequences of birthing routes such as Caesarean sections on infant health and our longer-term risk of chronic disease.

The problem with the hygiene hypothesis and the disappearing microbiome theory is that they don't fully explain why one child in a pair of genetically identical twins, delivered by the same route and raised in the same home, might develop an allergy when their sibling doesn't; they have the same parents, genes and, in theory, the same environmental exposures. A study in the United States of eighteen genetically identical twins, where one suffered from food allergy and the other didn't, shed some light on this conundrum. The twins with an allergy did indeed express significant differences in the diversity of their faecal microbiomes when compared to their identical siblings without an allergy. We don't fully understand why, but these changes were very persistent and continued into adult life and the difference may be something as seemingly insignificant as a single dose of antibiotics. Perhaps the more interesting observation was that the healthy twin microbiome was able to maintain the production of important metabolites that interact with the immune system.[13] This is significant because it means that the loss of specific bacterial species may be less important than the loss of their shared molecular functions. The functions themselves can be shared across different species or they can even be preserved within specific niches. This means that the microbiome's importance to our immune development cannot be explained by ecology alone, and the loss of its collective functions that can vary even between genetically identical individuals may be more important.

The timing and magnitude of exposure to allergens and environmental factors in early life are also vitally important. However, some scientists are starting to ask if these allergens are really sterile or if they have any microbial passengers. The Amish and the Hutterites (an Anabaptist German-speaking sect) are two US agricultural populations with similar traditional lifestyles who live in close but isolated communities.[14] Despite their common genetic ancestries, the prevalence of asthma and allergic sensitization has been found to be up to six times lower in the Amish than among the Hutterites.

The Amish follow strict traditional farming practices while the Hutterites use more industrialized ones, leading some scientists to suggest that it may be the environmental microbiome that causes this disparity in immune disease. They found that the microbial composition of dust samples from Amish and Hutterite homes, and the amount of toxins these bacteria produce, were 6.8 times higher in Amish homes. Amish children had significantly less activation of the innate immune cells that can cause the symptoms of asthma. When the scientific team then transplanted dust taken from these houses into an asthmatic mouse model, the microbes from Amish homes inhibited inflammation and the constriction of the airways in the mice, while microbes from the Hutterite homes did not. The implication is that the environmental microbiome also plays an important part in the development of our risk of allergies. It is therefore likely that urban microbiomes variably influence the development of our immune systems differently from, say, rural microbiomes. Urban microbiomes can be very discrete – for instance, they will still vary today between East and West Berlin.

While many of these hypotheses have merit, they also imply – dangerously, I believe – that some parents have a choice in the environmental exposures they allow their children to experience, or that living in dirtier homes or washing our hands less could be an aspiration. Killing pathogens that harm us is important and we still need to do this.

The fact is that something far more pervasive is happening. Environmental pressures like climate change, widespread antibiotic use, plastic exposure, fluorinated water consumption, dental fillings and a homogenized diet are important modifiers of the microbiome, and most of us simply cannot escape them. They may also be subtle, hard to measure and vary between twins. The sheer number of environmental influences on the microbiome is also massive, and because their impact occurs over multiple generations, they can be hard to spot and avoid. Blaser and Falkow's theorem is of exceptional importance to the future of human health and should be heeded by all health policymakers who wish to prevent chronic disease.

Alas, the microbiome was not studied during the fall of the Berlin Wall. Sadly, today we still have many other divided societies and cities for epidemiologists and microbiome scientists to study. As we retreat into our nation states at the beginning of the twenty-first century, walls are going up, not down. They may not be good for our society, but some scientists argue that the failure of the barriers within us that keep microbiota out are a syndrome of modern living, and that they should very much stay up.

The leaking gut

The gut barrier is a complex and wondrous thing. Like the Great Barrier Reef, it's a living ecosystem in its own right, composed of physical and immunological virtual barriers that protect the shoreline from waves and storms. The gut is designed to be damaged. It's lined with stem cells capable of regenerating all of the cells it needs to survive; this is just as well, because the cells that line the gut are regularly turned over, through a process of programmed cell death and through damage caused by food, debris and inflammation. Its regenerative properties are the reason the gut heals when I operate on it. All barriers leak to a degree, and those in the gut are designed to do so under specific circumstances and for specific symbiotic microbes with essential roles in maintaining our health. It's evolved to sustain a complex ecosystem of microbes, which in turn extends a wide influence over the innate immune response.

But the gut barrier also has to keep out dangerous storms in the form of pathogens and toxins. To do this, it secretes a thick wall of mucus that coats the billions of finger-like projections called villi that line the small bowel and their associated crypts, where our stem cells live. Mucus is like the atmosphere that protects the planet; it's made up of a subtle sub-layer and it's continuously renewed. This sophisticated and nurturing slime acts as a selective barrier that regulates nutrient diffusion and microbe transport to the underlying epithelial barrier, which is held together by tight junctions. The immune system still has to communicate with the bacteria that live in faeces through this barrier, and it can do so in a number of ways. Dendritic cells are cells of the innate immune system that have long arms capable of squeezing between these junctions – like an octopus squeezing through an impossibly small hole – to reach out and engulf microbes and their antigens even when there's no overt infection or inflammation. Microbes can also communicate with both the innate and the adaptive immune systems through pattern-recognition receptors that line the cells of the gut. These cells possess a vast number of metabolites and proteins that send chemical messages directly to our major organs and the brain. Microbes have no need to scale the wall if they wish to modify our health.

The gut wall possesses other secondary layers of defences that stop intruders seeping into the rich supply of blood vessels that are needed to maintain the health of the mucosa and to absorb nutrients. Both the innate and adaptive immune systems act as virtual walls that help to maintain the anatomical borders throughout our body. We each devote a substantial amount of energy to the production of about 3–5g of the secretory IgA each day for this purpose, and the amount of IgA produced in association with mucosal membranes is greater than all other types of antibodies combined. This is an important regulator of microbial populations that live in our gut, and it binds to both commensal symbionts and pathogens. Over hundreds of thousands of years, bacteria such as *Bacteroides fragilis* have learned to subvert this mechanism and change the expression of their surface antigens to encourage binding of IgA, so that they can safely navigate the immune system. The gut can also dynamically adjust its metabolism within the cells of this barrier to regulate local microbial populations – and if that doesn't work, it's able to produce natural antibiotics in the form of antimicrobial peptides (AMPs) for crowd control.

Human health depends on the health of these barriers, just as oceans depend on the health of their reefs. Defective epithelial barriers have been demonstrated in affected organs in asthma, atopic dermatitis, allergic rhinitis, chronic rhinosinusitis (CRS) and inflammatory conditions of the gut such as eosinophilic oesophagitis. That's because allergens, pathogens and environmental toxins can damage these epithelial barriers – for example, allergens from dust mites, toxins contained in laundry, dishwashing and household cleaning agents. What's more, surfactants, cigarette smoke, particulate matter, diesel exhaust, ozone, nanoparticles and microplastics all damage the delicate lining of our lungs and our gut. All of these were abundant in West Germany and became so in East Germany after reunification.

Whatever the environmental driver, failure of the gut barrier leads to 'translocation', or bacteria crossing the gut wall. Once on the other side, they are caught by macrophages, innate lymphoid cells and dendritic cells, ultimately switching on the innate and adaptive immune systems. Because the wall is injured, it releases molecules to alert the immune system for help, creating a continuous cycle of damage. As the wave of destruction builds, the gut becomes deprived of oxygen and the gut environment changes. This in turn benefits anaerobic pathogens that don't need oxygen to live – which bloom into an angry mob that can no longer be controlled. Symbiont populations get attacked or outcompeted, and they stop maintaining the health of the gut. Unless something happens to break this cycle, we experience symptoms or a disease. If very particular lymphocytes (known as helper T cells, or Th_2) are activated as part of this low-grade inflammation, it's more likely to cause conditions of allergy like asthma. If a subtly different type of helper lymphocyte (Th_1 cells) is activated, it causes chronic autoimmune diseases such as inflammatory bowel disease. A failure in the commensal intestinal barrier, which is maintained by symbiotic microbes, is therefore an important part of the story in the development of allergies. If it is broken, the body cannot educate the immune system to elicit an appropriate response when it meets a peanut for the first time. However, I must issue a word of caution.

The term 'leaky gut syndrome' is often used by health influencers to describe the importance of these barriers to our health, but I find this term misleading for three reasons. First, there is limited evidence that some people are particularly prone to the systematic destruction of the intestinal barrier and that this is the single unifying cause of specific immune diseases. Second, there are very few repeatable and reliable measures of intestinal permeability (or 'leakiness') that give us information on exactly which bit of the gut is leaking and by how much, or what or who is pouring through the defences of these failing barriers. Third, a major part of the gut defence is virtual; it's not an actual barrier, but a series of immune cells and antibodies. It's not the leaking that is important, it's the inflammatory response to the barrier injury, and this changes from person to person. The word 'syndrome' describes a constellation of symptoms caused by a single disease and there is no evidence that this is the case in the leaking gut. You should be suspicious of anyone who uses this term, not least because those who do so are often trying to sell you a cure, a diet or a test.

How I learned not to love the bomb

When the wall divided East and West Berlin, it created two geographically co-located compartments where the exposomes were radically different. When it was emphatically torn down, an entire population's microbiome had to adapt to a radically altered exposome, with degrees of individual variance. Yet the immunology bomb didn't immediately go off for those living in East Berlin when their countries were reunified. That is because it took a new generation of Berliners to be born into the new Western microbiome. The observations made in East and West Berlin only tell us so much. They don't reflect the broader influences on our exposome and internal ecology created through climate change and more modern forces of globalization since the fall of the Berlin Wall. A superheated microbiome war is now raging.

The power of the WOPR immune system is so great it's hard to regulate and control in a globalized culture that constantly antagonizes it. From an evolutionary perspective, we exist in a molecular nuclear arms race: the exposome, the microbiome and our immune dictators are continually raising the stakes, with each threatening to push the button every time they are displeased or threatened.

The modern inflammatory pandemic is happening because

of genome–exposome–microbiome (GEM) interactions that are defined at conception, and which persist until our death. Young people disproportionately experience the consequences of this perturbed relationship because of an immune system that is set to game this scenario until Matthew Broderick intervenes to save us. We still don't have the complete mechanistic picture that allows us to link specific bugs to specific food allergies or, much more importantly, the collective immune functions of these microbes across networks.

Many of the discoveries of our immune system were made in the decades after the Second World War, and scientists often relied on the language of the Cold War to describe its role in human health. The immune system has been depicted as a defensive system, just as I've described it here, primed to destroy attacking pathogenic organisms and to defend us at all costs, even if that means killing us through a process of mutually assured destruction. But the language of war is inadequate when describing microbiome–immune interactions. Our immune system doesn't just defend us; it is complicit in the lifelong evolution of our molecular identity and it is part of our environmental memory.

The cells of the immune system are therefore not simply killers, but sensing agents that collude with commensal microbes to promote health and wellness. The microbiome is hard-wired into our immunological operating system. The innate immune system promotes the growth of beneficial members of the microbiota and maintains a stable community of microorganisms. A healthy microbiome is simultaneously good and bad for us; it produces both pro- and anti-inflammatory toxins, proteins and metabolites, which shape the immune system over the course of our lives and across evolutionary time frames.

So, can the microbiome be leveraged to prevent allergies and diseases of the immune system? Data from faecal-transplant trials in animals and humans for conditions such as dermatitis are promising.[15] Researchers from the Food Allergy Program at Boston Children's Hospital have also treated both adults and children with severe peanut allergy using this approach. Four months after treatment with antibiotics, followed by FMT 'crapsules', children had increased their tolerance from 100 to 600 mg of peanut (about two peanuts) before having a reaction.[16] But these are early days, and these are small numbers.

There's still so much we don't know about how our microbes influence our immune system. Where we have gaps in the explanation of our suffering, we create stories and gods to fill them. Adventists – a group of Protestant churches that trace their origin to the United States in the mid-nineteenth century – often follow vegetarian diets, and numerous studies show they live longer and have lower risks for many cancers than the general population. This is not an act of a benevolent god; the miracle is explained by the gut microbiome metabolizing all of that lovely plant fibre.

A key doctrine of Buddhism is *Pratītyasamutpāda*; its basic principle is that all things arise in dependence of other things. According to Buddhism, the rebirth of sentient man depends on our understanding of this fact, and we will be unable to eradicate human suffering until we do so. It's a useful way of thinking about the microbiome–immune axis: a dynamic and constantly moving interconnected system where everything has a cause and an effect. Nothing is permanent and nothing is stable. Our immunology continually communicates with the exposome, and the microbiome, and it connects us to our ever-changing external and internal environment. We are one. A healthy microbiome requires balance, not bombs.

Namaste.

5. Sex and Bugs and Oestradiol

When I was fourteen, I met a girl. I had gone swimming with my brother at our local sailing club on the south coast of England. The aim was a lazy soak in a large blue steel testing tank built into the wooden decking of a jetty. Officially this was used for checking if sailing boats would float before they were sent out to race, and unofficially it was the village swimming pool. The tank was full of all manner of protozoa and plant life, but my brother and I could not have cared less; it was a cloudless hot afternoon at the beginning of August, the cool water was free from jellyfish, and it was clean enough that you could just see your feet if you were standing up in the shallow water. More importantly, my parents let us go on our own, and I had cigarettes. I had long hair and my choice of courting plumage was bright-red dungarees.

We weren't alone that afternoon; tens of other teenagers had descended on the pool. The girl in question seemed to have come from nowhere. I had never seen her before, and I was too awkward to go and ask where she had come from. Her eyes were electric blue, her hair was jet black and her swimsuit was Princess Di. This new girl and her family were impressively loud and, like the idiot teenage boy I was, I thought a water fight would impress her and, surprisingly, it seemed to work. The tide was high, and the warm sea breeze brought with it the promise of evening beach parties and kissing. There was no other option but to fall in love. I am still in love with that girl.

How we fall in love remains a blissful mystery. Chance is everything, and romance is predicated on being in the right place at the right time. Throughout the history of the human heart, scientists and artists have attempted to explain the chemistry that defines lust, love and attachment. More modern attempts to turn love into a marketable data science have explicitly aimed to subvert this chemistry for profit, by turning romance into a tedious, mechanized chore. Lizzie Bennet would not have coped on Bumble.

Evolutionary biology is even colder; it tells us that lust and love are merely selfish genes, socioeconomics, hormones and neuropsychobiology. My question is this: what if Romeo and Juliet only got together because their microbiomes ignited their romantic firework display? In other words, could their microbiota have influenced their ill-fated choice of partner and modified their disastrous mating behaviours? If nothing else, the microbiome is a by-product of our socioeconomic status and it certainly manipulates the physiology of our sex hormones. Despite the animosity between the Montagues and the Capulets, it is quite possible (in this fictional scenario) that their microbiomes were compatible, as they would both have grown up in the same environmental niche of Verona. Extroverted teenagers like Romeo and Juliet who have larger social networks also tend to have a more diverse gut microbiome, which may have a mating advantage.[1]

In nature, there is strong evidence for the importance of the microbiome to a successful courtship. Take the sexy fruit fly (*Drosophila melanogaster*), for example; prospective mates are chosen based on their choice of diet, and anyone who has gone on a date will know that food can be a potent aphrodisiac or a major turn-off. However, if the fruit-fly suitor is treated with a course of antibiotics, the mating preference conferred by the diet is abolished. No matter how much erotic fruit he eats, he will not get a date. If that same fly is then inoculated with bacteria from his food, his fly attractiveness is restored once more. In this experiment, this effect could be controlled by just one bacterium, *Lactobacillus plantarum*, which works by altering crucial chemical signals made of hydrocarbons excreted on the fly's armour.[2] So don't take antibiotics immediately before a date, and if you are taking them, maybe a date isn't the best idea anyway.

As per the Red Queen hypothesis (see page 25), it has long been known that both parasites and pathogens influence female mate choice. It has been argued that male mating plumage indicates parasite resistance, so females choosing males with more elaborate ornaments may gain genetic benefit because their offspring inherit parasite resistance. But this may also happen through a molecular 'landscape of disgust'. In other words, if it smells gross, don't touch it, and certainly don't kiss it. Female rats and mice prefer the urinary odours of healthy uninfected males compared to males experimentally infected with influenza virus, *Salmonella*, nematodes and protozoa.[3] Primates also use smell to assess genetic relatedness and to avoid inbreeding,[4] and they will avoid grooming group members infected by parasites transmitted by faeces.[5] Olfactory chemical communication influences other social processes, too – such as marking territory, distinguishing social groups and recognizing kin or potential mates. The anogenital gland secretions of the giant panda, for instance, are used to provide information on their age and sex to potential mates. These glands contain a diverse community of fermentative bacteria with enzymes that produce volatile odorants and the chemical code that permits this communication to take place.[6]

A human's olfactory signature is also composed of a unique combination of chemical odorants and pheromones in part produced by our microbiota. It's not clear how much control we really have over the make-up of our odour-producing microbiota. Pubescent teenagers typically reject all control (by which I mean soap) and as a result they have a very particular and pungent odour, defined in part by their unwashed skin microbiome. The sour odour of teenage armpits is the result of isovaleric and acetic acid produced by *Staphylococcus epidermidis* and the sulphur produced by the *Staphylococcus* species.[7] Teenage boys will often try to disguise this with dense clouds of deodorant and aftershave. This may not be the right strategy, because the human genome is well placed to take advantage of our natural chemical signals, with 300 gene clusters committed to olfaction.

When Romeo said, 'Did my heart love till now? Forswear it, sight, for I ne'er saw true beauty till this night', what he meant was: visual cues count. Anyone who uses a dating app knows this too, because we mostly post pictures in which we believe we look our best. This is tricky because attractiveness is a construct that varies with time, cultures, fashion and the whims of digital influencers. However, many of us do worry about how our skin looks, because it is an important visual cue of our health, and as a result our sexual and social confidence is often left at the mercy of spots and, in severe cases, *Acne vulgaris*. This is an inflammatory condition of the skin, which causes both physical disfigurement and psychological trauma. The causes of acne are multifactorial, but a common pathway occurs when a bacterium called *Cutibacterium acnes* blooms by feeding on the oil and sebum that are produced in hair follicles and sweat glands.[8] Subsequently there is a loss of microbial diversity, and particular strains of *C. acnes* dominate; this not only damages the skin, but takes the brake off the skin's immune system (just like in the gut), causing inflammation, redness and spots. It's not only bacteria that can influence the health of our skin, though. *Malassezia* is the most abundant fungus in the skin and it is also a cause of acne; it contains an enzyme called lipase, which is one hundredfold more active than the same enzyme found in *C. acnes*, which damages the skin, causing inflammation.[9]

We know that eating dairy products, refined carbohydrates, chocolate and saturated fats can make acne worse. This is because our skin's immune system is also influenced by the gut microbiome, and the gut microbiome of acne sufferers is less diverse than that of those who do not suffer from this condition. In particular, the bacterial phylum Actinobacteria are less abundant and Proteobacteria are more abundant. This is because species from these phyla are particularly adept at metabolizing fatty foods, and a by-product of this process are molecules that promote inflammation in the skin.[10] The variability in our gut microbiomes goes partway towards explaining why some of us respond to the antibiotics commonly used in its treatment and others don't.[11]

These variations in the gut microbiome are also gender-specific[12] and the gut microbiome's functions also change with surges in sex hormones during the menstrual cycle or puberty, which is in part why teenagers and young people are more commonly afflicted with spots.

After we'd known each other a few months, the girl wanted to hold my hand in public. I suggested that maybe things were moving too quickly and that we would both benefit from a temporary break. With that, she was gone and, just like Romeo, I was inconsolable. Could it have been my microbiome that made me say something quite so stupid? The microbiome subverts and influences our dating strategies because it contributes to our dynamic displays of courtship through microbial chemistry and the regulation of our immune system. But it also influences our fertility and our reproductive health and it explains in part why, had they lived, Romeo and Juliet would have experienced varying risks of chronic disease. We urgently need to understand how the sex-specific microbiome works to influence our health and well-being.

Sex

Microbiome science is not only providing a deeper understanding of how our sexual practices shape the populations of pathogenic and symbiotic bacteria that reside within our sexual organs, it is also providing new insights into how microbiota modify our sexual behaviours.

Compared to the gut or the lung, our urogenital anatomy is a relatively low-abundance niche for microbes. Despite this, our genital microbiota play an important role in maintaining the health and function of these organs, because they have co-evolved into that role. For example, semen typically has very low abundances of bacteria, such as those of the *Streptococcus*, *Corynebacterium* and *Staphylococcus* genera.[13] Its microbiome is transient because it is optimized for dispersal and colonization through sexual transmission. The penis is inhabited by diverse bacterial families, including the Corynebacteriaceae, Pseudomonadaceae and Oxalobacteraceae families and, as you may expect, these populations shift after circumcision.[14] A person's penile microbiome also has implications for their partner's chronic sexual health. For example, the microbiome composition of a man's penis predicts the risk of bacterial vaginosis, a common cause of vaginal discharge, in their female sex partner.[15] When and how women have sex after antibiotic treatment for bacterial vaginosis also influences their recovery, because penetrative sex affects the make-up of the vaginal microbiota.

A healthy vaginal microbiome is usually dominated by *Lactobacillus* spp. – hard-working species of bacteria that help form a physical barrier against pathogens and stimulate defences. *Lactobacillus crispatus*, *L. gasseri*, *L. jensenii* and *L. iners* are the dominant species, and among other things they help regulate the acidity (pH) of the vagina by producing lactic acid, which in turn helps maintain the health of the vaginal ecosystem.

Although the female genital tract is dominated by bacteria, other microbes are also crucial to its health. Yeasts are obviously important; 75 per cent of women will have at least one episode of candidiasis – or thrush – typically caused by the yeast *C. albicans*. Bacteria and yeasts in the vagina also coexist with viruses that help regulate these populations. For example, more than 400 phage have been identified for the thirty-nine *Gardnerella* strains of bacteria that are associated with the condition bacterial vaginosis, and almost 90 per cent of genomes of the vaginal and urinary *Gardnerella* strains contain at least one genome from a phage embedded in their bacterial DNA.[16] This means that phage play an extremely important role in the regulation of the vaginal microbiome and the maintenance of urogenital health.[17] So no matter what Gwyneth Paltrow or any other celebrities say, if you have a vagina, do not steam it. You need your urogenital symbionts to keep you healthy.

The act of sexual reproduction is not only an efficient method for transferring genes to our offspring, but is also a highly efficient practice for sharing microbes with a mate. Of course neither reproduction nor penetrative sex is necessary for this to occur. And, as we have discovered, kissing may also have had a protective and survival benefit for our Neanderthal cousins, because oral microbes are important for food digestion and for our dental hygiene. In ten seconds of intimate kissing you'll receive about eighty million bacteria from your partner.[18] Long-term partners have a similar tongue microbiome composition, but maintaining these colonies takes effort. You must kiss regularly, and the way you kiss makes a difference: a passionate snog will transfer more bacteria than a simple peck on the lips, so make time for it once in a while.

The problem is that we have become a little too efficient at transferring pathogens through this process. One in five people in the United States has a sexually transmitted infection (STI) on any given day, totalling nearly sixty-eight million estimated infections each year. The diagnosis and treatment of STIs is now a booming industry.[19] The majority of STIs are caused by just thirty bacteria, viruses and protozoa, so our understanding of the urogenital microbiome is skewed towards the small number of pathogens that have been widely studied. Today, the most common STIs encountered are the human papillomavirus (HPV) and the bacterial infections chlamydia and gonorrhoea. Almost all sexually active people will encounter the human papillomavirus at some point. These double-stranded DNA viruses are broadly classified into high-risk and low-risk, depending on their predilection to cause cancer. We now think of cervical cancer as sexually transmitted, as it's almost exclusively caused by the HPV virus, as are more than 90 per cent of anal cancers and a smaller proportion of vulvar, vaginal and penile cancers. In the UK, both male and female children are now vaccinated against this STI – a public health intervention that will save many lives.

Our sexual behaviours may have promoted STI virulence, but they have also influenced the evolution of our symbiont populations, and these populations exist in a steady state of competition that modifies our treatment response to STIs. HIV is a good example of this. In 2020 almost thirty-eight million people were known to be living with HIV globally, and until COVID-19, rates were dropping.[20] HIV appears to reduce the richness of the gut microbiome, and this loss of diversity is most pronounced in men who have the least response to antiviral treatment.[21] HIV infection is also associated with less diversity and richness of the semen microbiome, which interestingly can be restored after six months of antiretroviral therapy. Every week about 5,000 young women aged fifteen to twenty-four also become infected with HIV. It's more common for women with bacterial vaginosis or a high-diversity, pro-inflammatory vaginal microbiome that is low in abundances of *Lactobacilli* to acquire and transmit HIV (and herpes simplex virus-2 and HPV).[22] However, the precise strain of *Lactobacilli* is important; for example, *L. crispatus* is associated with reduced HIV risk while *L. iners* is not. Women with HIV also have more frequent and more resistant strains of *Candida albicans*. The biofilms of this yeast appear to harbour viruses such as HSV-1 where they lie dormant, protected from antiviral agents and the host's immune system.

STIs have had an evolutionary impact on our sexual behaviour, most commonly because they influence mate choice. Bizarrely, in some species, such as the cricket *Gryllus texensis*, STIs increase sexual activity among males. But mammals typically want to avoid reproduction carrying a sexually transmitted pathogen, and we subconsciously look for signs that suggest a potential mate has a harmful microbiome. During the eighteenth century, when many STIs either had no cure or were treated with highly toxic mercury, this signalling was overt: fake moles were created in response to the outbreak of smallpox and were adopted to cover the marks of sexually transmitted diseases such as syphilis, caused by a spiral-shaped bacterium called *Treponema pallidum*. The French called these facial decorations *mouches*, meaning 'flies'. Black spots also famously denoted syphilis on the faces of the greedy and squalid aristocracy (and physicians) in Hogarth's satirical caricatures *Marriage A-la-Mode*, painted in 1743.

Our sexual practices and sexual signals are continuously adapting, driven in part by an evolving digital culture. TikTok videos don't come with a digital *mouche*, largely because stigmatizing those with STIs would be harmful. But digital dating apps do allow us to build sexual networks that will impact on the evolution of the human microbiome, and the only way to really know if a potential partner has an STI is to get them checked. I would highly recommend regular sexual checks-up if you're having sex, no matter your age, gender or sexual orientation.

It seems very likely that our society's shifting attitude to dating, and our use of digital technologies for this purpose, are influencing the stability of the symbiotic sexual microbiome in a similar manner to STIs. But the scale of this change and its consequences are not yet known. I would argue that this now needs to be urgently examined, not only because it influences our risk of suffering from chronic disease, but also because it promotes sexual health.

Beyond the genderome

A lot of the differences in the microbiome that exist between sexes stem from variations in our *reproductive* anatomy and physiology, which to a degree defines the ecology of these microbial systems. Our reproductive health also demands a specific set of microbial functions, and in return we provide a specific set of environmental conditions that allow our microbes to survive – which means this is a two-way conversation. However, the male and female microbiome is also greatly influenced by the environment, and the *productive* exposome varies greatly between the sexes. This means that our sex influences where and how we work, our roles in our families and our societies. So it's possible that our microbiome characteristics are just as much the result of our exposome or environment as they are of our physiology. These differences in the exposome also play a major role in explaining variations in health outcomes between sexes and genders. The sum of these environmental and genomic functions of the reproductive microbiome is often referred to as the 'genderome'. This concept may explain the biological and health differences that we observe between individuals of different sex in the same species, known as the 'gender dimorphism'.

For example, in the developed world, women who are female at birth typically live longer than their biological male equivalents and are less likely than men to experience a heart attack, cancer (excluding those found in sex organs), alcoholism, cirrhosis, Parkinson's, schizophrenia, Autism Spectrum Disorders (even when accounting for diagnostic bias)[23, 24] or substance abuse. Because of socioeconomic and gender-based social inequalities that exacerbate any underlying genetic variance that exists between the sexes, women also experience higher rates of obesity, stroke, gallstones, autoimmune disease, osteoporosis, Alzheimer's and eating disorders. For girls aged fifteen to nineteen, self-harm is the second leading cause of death, globally, after pregnancy.[25] In the developing world, millions of women still don't have access to basic health services and modern contraceptives. As a result, the burden of communicable diseases and perinatal and nutritional disorders remains unequally distributed between the sexes. Transgender people also face a disproportionate risk of death – double that of cisgender people – in large part because of a lack of access to adequate healthcare and high rates of violence and discrimination.

So what role do microbes actually have in determining our sex, and can they really explain this dimorphism? Some insects and nematodes abdicate all responsibility in determining their sex to what I would describe as 'transbacteria'. Probably the most famous gender-transforming bacterium is *Wolbachia pipientis*, which manipulates its insect hosts by selecting the gender that best suits its own selfish survival requirements. It can even trigger the spontaneous development of an embryo from an unfertilized egg (known as parthenogenesis), or simply render its host infertile by preventing sperm and eggs from forming viable offspring in a process known as cytoplasmic incompatibility.[26]

When it comes to humans – as far as we're aware – microbes don't play a role in determining sex. But they most certainly influence the health consequences of our sex, and this occurs from the moment we're conceived. For instance, male mice that are born without any contact with bacteria – known as germ-free mice – develop feminine traits, while germ-free females become masculinized. This happens because bacteria in the gut influence how growth hormones work during development and sexual maturation. Our microbiome is important for ensuring that sex hormones are made correctly in the liver and in our fat cells.[27]

The gut microbiome also has a sex- and time-specific impact on our brain development. For example, microglia are immune cells in the central nervous system that influence brain health. If the microbiome is depleted in early life, for example through antibiotic use, this in turn alters the function of microglial cells, but the health consequences of this disruption are experienced differently in men and women.[28] This explains why rodents of different sexes given antibiotics experience pain variably, and the development of sex-dependent changes in brain function. The implication of this work is that gut microbiome–immune interactions may explain human susceptibility to sex-dependent disorders of mental health, such as depression, and neurological degeneration, such as Alzheimer's.

It follows, then, that taking antibiotics before puberty influences our sexual phenotype, and in turn the microbiome determines how different sexes respond to antibiotic use. The impact of exposure to antibiotics on newborns was studied in 12,422 children born at full term from the Southwest Finland Birth Cohort. The research showed that early antibiotic exposure had a long-term impact on the gut microbiome that resulted in reduced growth in boys during the first six years of life, but not in girls. And antibiotic use later in childhood was associated with increased Body Mass Index in both males and females.[29] So boys and girls have different vulnerabilities to chronic diseases that are determined by the microbiome before puberty, and these differences may last for the duration of their lives.

Everywhere in the world, and across all ages, the human gut microbiome is different in males and females. In research led by the Human Microbiome Project, 242 healthy male and female subjects had their microbiomes sequenced from fifteen different body sites in the airway, mouth, gut and skin. Seven out of fifteen sites exhibited significant sex differences, although some of these were only detected at phylum level (the broadest possible taxonomic rank).[30] The skin showed the greatest differences between sexes, but the microbiomes found in all body sites were highly sex-specific.

This observation was corroborated in an analysis of 2,338 adults from the Pinggu (PG) district of Beijing, which revealed a core set of ten microbial groups enriched in women that were also seen in other female participants from China, Israel and the Netherlands.[31] These observations were still present when genetic, dietary and other clinical variables were accounted for. The differences were most pronounced in middle-aged adults, before gradually decreasing with advancing age and the menopause, and women had a much more diverse gut microbiome than men. However, these sex-specific bacteria were undetectable in prepubertal Dutch girls aged six to nine, suggesting that these changes were strongly under the influence of sex hormones. Many other studies have found similarly strong associations between the presence of specific gut bacteria and sex hormones. For example, *Prevotella* has a strong positive correlation with testosterone, and negative associations with oestradiol. To understand the implications of this observation we need to understand precisely how the microbiome actually interacts with sex hormones.

Sex talk – hormones and microbes

Oestrogen, progesterone and testosterone are sex hormones that are collectively known as androgens. The honesty of our sexual signalling is strongly mediated by androgens (just ask any teenager), and these would most certainly have been raging when Romeo and Juliet first met. But these hormones also play an important role in organ development and in the maintenance of our day-to-day reproductive health through a well-established series of molecular pathways.

Oestrogen is a wonder-hormone that influences every aspect of reproductive health in women (and it's also important in men). The microbiome is so intimately involved in the function of oestrogens that we've recently coined the concept of the 'oestrobolome' – the total collection of bacterial genes that break down (or metabolize) oestrogens. This is so important because bacteria in the gut are able to influence how much active oestrogen circulates around the body. They can achieve this effect because they possess enzymes that change the chemical structure of oestrogen by unbinding it from bile acids or proteins. Once free, these oestrogen molecules can be more easily reabsorbed in the intestine and returned into the circulation, where they signal through oestrogen receptors and exert their biological effect.

The ovaries are not the only organ that produces oestrogen.

Specialist cells found in the gut also synthesize a much smaller-quantity hormone called oestradiol – and this process, too, is influenced by the microbiome. This is important because this hormone has specific functions within the gut. For example, both innate and adaptive immune systems have receptors for sex hormones, which direct the maintenance of intestinal barriers, and progesterone even promotes the production of IgA, which in turn reduces permeability to bacteria and toxins in the gut. The link between the gut and sex hormones is so important that circulating levels of immunoglobulins in the blood fluctuate, depending on the stage of a woman's menstrual cycle. Oestrogen also acts as a vasoactive hormone, increasing local blood flow, while supporting the collagen content of the vagina and producing lubricants. The benefits of oestrogen are seen in the bladder, urethra, pelvic-floor musculature and wherever else oestrogen receptors are found.

The interdependency between androgen function and the microbiome means that the functions of a woman's microbiomes are also influenced by her menstrual status and cycle, and this relationship fluctuates over varying timescales. And because the microbiome plays a role in defining oestrogen bioavailability, it can in theory influence all of these bodily functions. For example, the microbiome plays an important role in defining, via the gut–brain axis, the symptoms that many women experience before or during their periods, such as bloating, pain or feelings of anxiety. It is also becoming more widely accepted that because the microbiome interferes with the bioavailability of all androgens in a woman's body, it may also play an indirect role in the cause of chronic diseases associated with their regulation – for example, endometriosis and polycystic ovary syndrome (PCOS), which affect up to 10 per cent of women and significantly reduce health, fertility and quality of life in these individuals. It also goes a long way towards explaining much of the sex dimorphism in chronic disease, because the interaction between oestrogen and the microbiome potentially changes a woman's risk of obesity, metabolic syndrome, cardiovascular disease and breast cancer (men's risk of prostate cancer is also influenced by oestrogen). The microbiome now represents a new target through which these conditions can be risk stratified and prevented.

In the late 1940s, Dr Carl Djerassi began to synthesize progestogen from wild yams, leading to a series of discoveries that would ultimately translate into the contraceptive pill. It was approved for release in the US in 1960 and within two years was being used by 1.2 million American women. In the UK today, approximately 26 per cent of women aged sixteen to forty-nine use hormonal contraception and increasing numbers of women take the progesterone-only pill, which is also co-metabolized by the microbiome.[32] Despite the wide-scale adoption of this revolutionary medicine, relatively little was known about its interaction with the human microbiome. Based on small numbers of studies, the combined oral contraceptive pill does not appear to cause changes to the diversity of the gut microbiome. However, the pill can alter the gut microbiome's metabolic functions; an analysis of ten healthy pre-menopausal women taking the oral contraceptive pill at the Massachusetts General Hospital demonstrated that over a period of six months, levels of the sex hormone oestradiol in the blood were positively associated with relative abundances of specific microbes in the gut such as *Eubacterium ramulus*.[33] In addition, microbes that metabolize some specific types of amino acid were significantly associated with lower oestradiol and higher levels of total testosterone in the blood.

The impact of the gut microbiome on pill effectiveness may be even more important in obese women or in those suffering from associated chronic health conditions of the ovaries, such as PCOS, where the ability of the gut microbiome to co-metabolize androgens is already altered.[34] The pill also has a marked impact on the vaginal microbiome, and some analyses suggest it could reduce bacterial biodiversity, promoting cervical inflammation and in turn the risk of HIV transmission.[35] We therefore need to more urgently understand how the gut and urogenital microbiome are connected, so that we can both improve the safety of our contraceptive strategies and reduce the risk of STI transmission.

When the menstrual cycle stops, a woman's genderome continues to influence her health. Post-menopausal women have lower levels of *Lactobacillus* in the vagina, but more *Prevotella*, *E. coli*, *Streptococcus* and *Bacillus*. Post-menopausal women with a vaginal microbiome that is high in *Lactobacillus* tend to have fewer symptoms of genitourinary syndrome of menopause (GSM) – such as vaginal dryness, burning and irritation, or sexual symptoms like a lack of lubrication, discomfort or pain. We don't yet completely understand the relationship between the microbiome and the symptoms that come with a loss of oestrogen and progesterone. But many women take hormone replacement therapy (HRT) to treat hot flushes, night sweats, mood swings, GSM and loss of libido and, given the data on the contraceptive pill, microbiome–HRT interactions are likely to be important.

While HRT is generally safe, the microbiome plays a role in determining both its effectiveness and its side-effects. Evidence for this comes from studies where mice were given long courses of oestrogen-replacement therapy after their ovaries were removed. The researchers observed that HRT led to a reduction in the number of bacteria that produce the enzyme β-glucuronidase within the gut, which helps promote the hormone's bioavailability.[36] Long-term therapy with oestrogen changes the make-up of the microbiome, which in turn changes how women's bodies metabolize the very hormones it's being treated with.

Men are just as susceptible to gut microbial interference in their sex hormones. Scientists from Wuhan University in China conducted an experiment that demonstrated a correlation between depression and the presence of a particular bacteria that lowers testosterone levels. They isolated the bacterium *Mycobacterium neoaurum* from the faecal samples of testosterone-deficient patients with depression. This bacterium was able to break down testosterone in the lab.[37] When the gene for this enzyme was switched on in *E. coli* and fed to rats, it reduced the levels of testosterone circulating in the brain and in the blood, which in turn was linked with a change in behaviour associated with depression. In a sample of 107 men with depression, 43 per cent had this testosterone-busting enzyme present in their faeces, compared to 16.7 per cent in the control group.

As in women, obesity and insulin resistance in men also play a role in this dynamic, and men with type II diabetes and testosterone deficiency have different gut microbiota compositions from those without diabetes.[38] Not only does this have important implications for their risk of chronic diseases like cancer, but the future of our species may depend on it.

A fertile hunting ground

The total fertility rate measures the average number of children per woman, and globally this has steadily fallen over the last fifty years from 4.5 to 2.3 children per woman. The global fertility rates are expected to continue to decline over the next few decades, with dramatic consequences for our species and our ageing society.[39] It's well established that sexually transmitted pathogens, such as *Chlamydia trachomatis* and *Neisseria gonorrhoeae*, are a major cause of female infertility. However, the precipitous decline in human fertility can't be explained by pathogens alone, and this is not a problem of just one sex.

The average sperm concentration fell from an estimated 101.2 million per ml to 49.0m per ml between 1973 and 2018 – a drop of 51.6 per cent.[40] The gut–testis axis may explain some of this observation. First, we know that obesity, insulin resistance and a high-fat diet all change male reproductive function and the production of sperm (known as spermatogenesis). This is significant because we are in the middle of an obesity pandemic. Animal models may explain why: sheep fed with a high-fat Western diet develop a metabolic syndrome of obesity and insulin resistance. This in turn influences the absorption of vitamin A, which in turn contributes to abnormal spermatogenesis. Intriguingly, it was possible to replicate this metabolically derived infertility through faecal microbiota transplant to animals without metabolic syndrome.[41]

However, it takes two to tango, and reproduction ultimately requires hard-swimming sperm to reach and then fertilize an egg. To complete this epic journey, the sperm has to make it all the way through the female genital tract and its defence systems. For this reason, the vaginal, cervical and endometrial microbiomes have also been studied in infertility, and it does appear that they play an important role in determining how sperm reach their target. For now, studies of diversity and species richness among infertile women are quite often conflicting, and the precise determinants of the optimally fertile vaginal microbiome remain unclear. We are, though, starting to gain some insights.

For example, during the artificially induced ovulation cycle of *in vitro* fertilization (IVF) treatment, women experience fluctuations in vaginal microbiome diversity, yet this appears to be independent of the gonadotropin-releasing hormone (GnRH) hormone, which is used to 'switch off' the brain from stimulating the ovary.[42] This means that despite the strong relationship that oestrogen and progesterone have with the vaginal microbiome, there may be hormone-independent mechanisms that regulate a woman's vaginal microbiome to ensure it is optimized for reproduction.

Women who have IVF treatment for infertility tend to have lower vaginal *Lactobacillus* levels than fertile women, and the same is true for women who go into pre-term labour. This is a little counter-intuitive, if true, as the vaginal probiotic species *L. crispatus* significantly reduces sperm motility and sperm penetration. A study of 192 women who underwent a fresh embryo transfer found that implantation failure could be correctly predicted in thirty-two out of thirty-four women, based on their vaginal microbiota composition. The presence and degree of dominance of *L. crispatus* was an important factor in predicting pregnancy.[43]

As with anything else in the microbiome sciences, it is likely that timing and context are everything in explaining these observations. First, the precise anatomical location of the *Lactobacillus* niche is important. For example, the presence of *Lactobacillus* spp. in ovarian follicular fluids has been associated with higher rates of embryo transfer, and improved pregnancy outcomes in IVF procedures, in both fertile and infertile women. Second, it may be the case that *L. crispatus* has a protective effect on maternal health and offspring development in fertile couples, where its presence prevents the combination of abnormal sperm and eggs. However, higher abundances become problematic in males with low sperm counts or defective sperm.

It is more probable that fertility is defined by the compatibility of an individual's semen and the female genital-tract microbiome, which in turn defines the immune response of the female immune system to the process. The vaginal immune system, primed by its local microbiome, may be inadvertently targeting either sperm or their microbial passengers, inactivating them or causing clumping (known as agglutination), rendering the sperm useless. Microbiome compatibility between parents may therefore be a more important evolutionary determinant of mate selection and reproduction than previously appreciated.

The modern microbiome is directly and indirectly influencing our ability to reproduce – and it may yet be our most important target, when it comes to correcting the precipitous fall in industrialized populations.[44] The first step towards reversing this effect may be to improve our gut health and, more specifically, to address the obesity epidemic. The microbiome may yet serve as a therapeutic avenue for couples struggling to conceive.

Numerous studies of probiotics to modify the vaginal microbiome and make it more sympathetic to exhausted sperm are ongoing, but that is just one strategy. Male mice with type I diabetes fed on a high-fat diet produce defective sperm. When these animals were treated with an FMT augmented with a natural polysaccharide found in brown seaweed called alginate oligosaccharide (AOS), their sperm motility significantly improved.[45] We don't know yet whether FMT will translate successfully into human trials of infertility, but it's only a matter of time before someone tries.

The holosexual human

Sex and gender are not the same thing and, despite its importance, the microbiome does not define either of these biological or social constructs. The human microbiome is neither male nor female, and by this I mean that prokaryotes do not sexually reproduce and they should not be anthropomorphized to suit a specific agenda. If anything, microbes decided it was far more efficient to avoid pronouns in their asexual pursuit of reproductive efficiency. From an evolutionary perspective, the microbiome is dependent on the holobiont's reproductive cycle for its own survival. The microbiome is therefore not a dominatrix submitting us to its perverse parasitic peccadilloes; it's our life partner, nudging us towards procreation and promoting health that is both sex- and gender-specific, because it's in its selfish interest to do so. The microbiome contributes to the effectiveness of our sex hormones, and influences our gender-specific organ development and urogenital health. The gut microbiome also influences our mating rituals, from how we groom, look and smell, to our social behaviour through mate selection. Its structure and function are, in turn, influenced by our sexual practices. It is deeply entwined in both the sex- and gender-associated risks of chronic disease, wellness and fertility. I think of humans simply as 'holosexuals', loosely defined by the sum of our very sexy microbial genomes.

In practice, the prevalence of sexual disease isn't spread equally across the full spectrum of genders, and those who identify as LGBTQ+ experience a disproportionate burden. There are, of course, far more important social, behavioural and political obstacles to gender equality and the eradication of homophobia than microbes. But the microbiome can be leveraged to reduce gender dimorphism, and to cut the risk of diseases within and across social groups and sexual identities. A healthy urogenital and gut microbiome should be a legitimate and objective measure of sexual health, particularly as it varies according to the sexual orientation of the individual. But we won't be able to use the potential of the new science of the holosexual microbiome if we continue to exclude selective LGBTQ+ groups from this research – or prevent some members from having access to a resilient internal ecosystem. The microbiome should be incorporated into the ongoing social commentary on changing sexual and gender norms among young and old alike; and it should be leveraged by policymakers to make sure that the most vulnerable are protected.

A holosexual health strategy must also ensure the health of the genderome across all niches of the body, and across all stages of a holobiont's sexual development. So as our lifeline fluctuates with the monthly cycle of menstruation or the more permanent shifts of the menopause, pregnancy, an enlarged prostate or a midlife crisis, it is optimized for happiness, and not just for preventing treatment of disease. The most important time for achieving the health of the holosexual may well be before puberty, when the microbiome is at its most pliable and we can target its assembly or functions to benefit the individual. The implications of not doing this might be a lifetime of chronic disease risk.

Of course none of this has anything to do with love – maybe it's better we never understand that particular chemical formula, and I will leave this to Shakespeare. Whatever your gender or sexual choices, a happy relationship will ultimately come down to persistence, humour and forgiveness . . . and none of these factors are related to the microbiome. They are distinctly human traits. All I know is that I am incredibly lucky: the girl forgave me and today we still hold hands.

6. The Big Bang

On 8 September 2008, at 9 a.m., the Girl lay on an operating table. Her swollen abdomen was painted with a pink antiseptic wash and she was covered in stiff blue sterile drapes. Despite my familiarity with the operative process, I was ushered behind a temporary drape blocking her view of the operative field, and she told me in no uncertain terms to hold her hand. An obstetrician wearing a mask, a sterile gown and gloves spoke in unintelligible tones to her assistant as she made a low incision across the bottom of the Girl's abdomen. Within a matter of minutes, the obstetrician pulled out a pink baby boy, and held him upside down. Jacomo Kinross was very displeased and he reeled from the bright lights, the cold and the crowd. He was carefully checked by a paediatrician, warmed, cleaned and gently placed naked on his mother's breast. The great colonization had begun.

In some women, this occurs before birth; vertical transmission of bacteria describes the process through which microbiota are passed directly from mother to baby through the blood during gestation. In obstetric units across the world, midwives and doctors screen for pathogens that are commonly transferred in this way, such as rubella, HIV, syphilis and toxoplasmosis, as they cause tremendous harm to the unborn baby. But we may have been so focused on preventing the harm caused by these pathogens that we failed to notice the equally important role of maternal microbes that ensure the health of new mothers and infants alike.

In fact the symbiotic microbes gifted to you by your parents during your arrival in the world are the most important present they will ever give you. These microscopic life forms will educate and protect you and extend the power of your genes. Just as we have done with our planet, if we want to understand the importance of the microbiome to human health, we have to go right back to its inception and examine its evolution in our early life. From this viewpoint, the microbiome has an irreplaceable and irreversible influence on our organs.

The maternal microbiome

A mother's microbiome has superpowers that protect everyone involved in the reproductive cycle. Yet classical medical doctrine teaches us that we're delivered into this world from a 'sterile womb' – and this means there is not a single microbial life form within us. From an evolutionary perspective, it is clearly a good thing for any developing neonate to be kept well away from harmful pathogens. And so the amniotic sac and the placenta in which we all gestate act as an immune bubble, designed specifically to keep bad bugs out. While the placenta can temporarily harbour bacteria (for example, the bacterium *Streptococcus agalactiae* – a rare cause of infections in newborn children[1]), it doesn't have its own microbiome.

Microbiome science does, however, challenge the sterile-womb hypothesis. It points out the many ways through which maternal microbes are able to communicate with the developing foetus. The largest collection of bacteria that a new mother possesses is found within her intestine. This colossal metabolic turbine possesses a stupendous portfolio of wonderful functions that are severely tested during the gestation, birth and nursing of a child. And so it changes with the mother to help her fulfil these needs.

When a woman becomes pregnant, her gut microbiome stays the same for the first trimester – dominated by the Firmicutes and Bacteroidetes phyla. But by the third trimester, her microbiome has changed dramatically. The overall biodiversity of the gut gets streamlined as it focuses on the huge task of producing the energy needed to grow a new human,[2] although marked microbial diversity exists between pregnant women, and a pregnant woman's microbiome is unique to her.

We can think of the microbiome as an orchestra. The music it plays is eternal, and the sheet music is passed from mother to baby for generations. The mother's microbiome produces hundreds of thousands of different types of small molecular notes known as metabolites, lipids and large molecules called proteins that reach the baby through the blood. The unborn baby listens and adapts in response to these molecular signals during gestation until it is delivered and colonized with its own microbial musicians.

The tone, pitch and volume of its song change along our entire lifespan, and during pregnancy the orchestra builds slowly across the trimesters, leading to a dramatic 'ode to joy' right before birth. To achieve this musical goal, the maternal microbiome doesn't throw out the old orchestra or bring in a new one. Instead, it wakes up sleeping sections of the band and instructs them to play all the right maternal notes in the right order. Even after birth, the maternal microbiome continues to play to its offspring through breastfeeding, touch and play. The maestro is, of course, mum, and she conducts the orchestra through her diet, lifestyle and biological functions – some of which are outside her conscious control.

Pregnant women are often told they are eating for two. This tired cliché is incorrect: they are eating for trillions. This may, in part, explain why new mothers experience urges for foods that they wouldn't typically eat. The microbes in the maternal gut are dependent on her diet for their growth, just as much as a baby is, and they can only produce secondary metabolites of benefit for all parties if they're provided with the correct fuel. The microbiome subtly alters maternal sensations of satiety, hunger and food cravings to get what it needs. As the maternal diet changes throughout pregnancy, so the total microbial numbers in the oral microbiome also ramp up, changing how expectant mothers experience food.

All pregnant women are told to take the essential B vitamin folate, which is used by the body to make DNA and RNA. 'Essential' means that humans cannot produce it and we depend on consuming folate from our foods, such as green leafy vegetables, fruits, cereals and liver products. If a new mother has critically low levels in their blood in early pregnancy, it can cause neural-tube defects, or spina bifida, in the newborn. Folate is important to many other aspects of our health, and it's so lacking in Western food products that we've started artificially fortifying food with it. But folate is also manufactured by bacteria in the gut, providing us with up to 20 per cent of our recommended daily allowance.[3]

The microbiome doesn't only provide vitamins, but also essential amino acids and tens of thousands of different species of metabolites broken down from indigestible foods such as fibre. It detoxifies synthetic chemicals in our food, medicines and environments, which are usually foreign to the body, called xenobiotics. The microbiome breaks down all of these molecules for the selfish purpose of freeing up carbon that it needs for energy, and it has a large repertoire of enzymes with which to do this. The by-products of this process are the goodies (and baddies) that influence gut health. It's a central hub of a massive biochemical superhighway, or a grand central station sending its bursting locomotives to all corners of the body.

When an unborn baby is hungry it will demand feeding, but the only option for the gestating baby is the mother's menu. The meal is filtered through the mother's gut, the placenta and the baby's liver, and all three act as a metabolic defence against environmental toxins. Once filtered, the fuel can be piped out to the baby's rapidly growing and very hungry organs. These metabolites also act as signals that can influence the development of an unborn baby's immune system. The fact that a uterus lacks a microbiome may be unimportant, as the maternal microbiome is able to whisper to its baby through the placental wall.

When the maternal microbiome is changed because of her own health-related problems – for example, obesity – there's a higher risk of non-communicable diseases in the infant, and this risk may be lifelong. Overweight or obese pregnant women show fewer *Bifidobacterium* and *Bacteroides*, and an increase in some specific species of Firmicutes. These changes resemble those reported in non-pregnant individuals with obesity and the commonality between the conditions is striking, given the physiological demands and dramatic change of lifestyle that most new mothers experience. These particular bacteria are highly efficient at harvesting energy from a high-fat diet, and the faecal microbiome in the third trimester of pregnancy shows the strongest signs of causing inflammation in the gut, which could explain some of the changes we see in maternal health.[4]

For example, the microbiome has been heavily implicated in having a mechanistic role in the metabolic complications of pregnancy such as gestational diabetes, a temporary state that sees the mother's blood-sugar levels rise. This is screened for in new mothers because it can cause potential harm to the unborn baby. Scientists have explored the connection between the microbiome and gestational diabetes in experiments with mice and have found compelling evidence for the role played by the microbiome in this process. When faeces taken from pregnant women in the third trimester was transferred to a mouse, the researchers observed dramatic results. The new microbiota in the mice intestines triggered the production of fat cells in the mice and made them far less sensitive to insulin, the hormone that regulates blood sugar. The changes to the diversity of the gestating mother's gut microbiome that contribute to gestational diabetes are also surprisingly persistent and can last for up to eight months after the baby is born, which might explain why the condition persists in some women after birth.[5]

And this is not the only isolated example of how the microbiome influences maternal health. Similar changes in the diversity of the maternal gut microbiome seen in mothers with gestational diabetes have also been observed in pre-eclampsia, a temporary state of high blood pressure seen in pregnancy.[6] This implies that the maternal microbiome may have multiple influences on maternal health through shared pathways. We are now starting to understand quite how influential the maternal microbiome is on the health of the gestating baby, and there is no considerable interest in defining just when and how new mothers share their symbiotic microbes with their offspring.

The foetal microbiome

When a baby is born, its microbiota is very similar in make-up across every part of its body – with the notable exception of its gut.[7]

Its first poo is called meconium and it is a thick, green tar-like mixture of material ingested by the infant during its time in the uterus. It's made up of cells that line the bowel wall, fine hair, mucus, bile, amniotic fluid and – possibly – bacteria. The meconium shows strong correlations with the microbiota of the mother's faeces within twenty-four hours of birth.[8] If the meconium of an unborn baby isn't sterile, it begs the question: when precisely is it colonized – and by what or whom?

Before its heart starts to beat or its limbs begin to bud, an embryo's gastrointestinal system forms from a single hollow gut tube when it's just three weeks old. By week twenty-four, a foetus develops the ability to absorb nutrients through its intestine and technically, then, it's 'possible' that microbial life forms could live in the foetus's gut *in utero*. The innate immune system is the first line of defence in the developing baby, and it appears at about eleven weeks, with its varying cellular components emerging at different rates. The lymphocytes of the adaptive system (B and T cells) can be found in the developing intestine by about sixteen weeks of gestation; and by about nineteen weeks they are organized into specialized lymph nodes. By the time of birth, new humans have a fully functioning gut and an immune system, which is like buying a computer with pre-loaded antiviral software that works from the very moment it is switched on, although it does need a system update so that it can determine which viruses are harmful and which ones it needs for survival. Or does it?

Any bacteria willing to live in the nutrient desert of the developing gut while being continuously abused by the hormones of pregnancy, and attacked by the maternal and infant cellular defence systems, would need to be hardy. This would also be a tightrope to walk, because if it inadvertently overexcited the immune system, it could also harm the developing baby. Extremophile archaea, accustomed to living on limited resources on hydrothermal vents at the bottom of the Pacific Ocean, could potentially be up to the task of living in such a low-nutrient setting. So the idea that the developing human gut might be sparsely populated *in utero* with a very low abundance of microbiota such as the *Micrococcus* genus is hugely controversial.[9] These hardy bacteria are commonly found on the skin, but also in soil, water and dust, because they can survive in environments with little water or high salt concentrations.

If these microbiota really do exist in the developing gut, it is likely they originate in the maternal vaginal microbiome, although their precise role (if any) in educating the memory of the neonatal immune system remains to be defined. Data from experiments performed on gestating lambs suggest that the faint metabolic signatures of bacteria identified within the gestating foetal gut can be detected in the faeces of these lambs at birth and in blood sampled from the umbilical cord. This implies that these bacteria have the potential to chemically signal to the unborn child and directly influence its development. More worryingly, these experiments also suggest that viruses and bacteria identified in the gestating gut can carry antibiotic resistance genes, which can be transmitted from the mother to the foetal gut microbiome.[10] And remember that the role of the microbiome dark matter is completely unknown in this process.

If it is definitively proven that the developing gut has its own microbiome *in utero*, then the book of immunology may have to be rewritten. It would imply that bacteria and microbiota have a far more critical role in programming our immune systems, and that horizontal transmission of bacterial genes – that is, from bug to bug – *in utero* could shape our development. The gut–maternal microbiome–immune axis is, in my opinion, the most important area of focus when it comes to preventing all chronic disease. Defining exactly how this dynamic marriage develops is going to be a Nobel Prize for someone.

The birth of the human microbiome

At some point (nine months, give or take) every baby must be born. For now, it's generally considered that the baby's colonization begins with delivery through the birth canal, when we are first anointed with our mother's microbes. The maternal vagina possesses its own distinct microbial ecosystem that evolves during pregnancy. By the third trimester there's an increase in *Lactobacillus* spp. and a reduction in anaerobic bacteria that don't like the presence of oxygen. This change is important because it protects the baby from harm during delivery. Mothers with *Lactobacillus*-dominated microbiomes in the first trimester have a lower risk of giving birth prematurely, because *Lactobacillus* bacteria compete with pathogenic microbes and stop them growing.[11] What constitutes a healthy maternal vaginal microbiome varies by culture and geographical location. For example, variations in diversity have been seen in Latin American and Peruvian women,[12] Australian Aboriginals[13] and African American women. However, a common finding among these groups is that low abundances of *Lactobacillus* associate closely with pre-term delivery rates. This may well become the key to understanding global variations in infant mortality from pre-term births.

Many scientists are now asking whether the way we deliver our babies has an important impact on how bacteria are seeded from mothers to babies, and whether this, in turn, has implications for their longer-term health. A Caesarean section can be a lifesaving intervention, and it's the route through which both of my children entered the world. Don't waste time feeling guilty if you've had one; there are plenty of benefits, and you're in good company. The global rate of Caesarean birth has doubled in the past fifteen years to 21 per cent, and it's increasing each year by 4 per cent. The likelihood of a baby being delivered this way depends on what part of the world they're born in. In southern Africa fewer than 5 per cent of babies are born by C-section, while the rate is almost 60 per cent in some parts of Latin America, including Brazil.[14]

The gut microbes of babies delivered by Caesarean section are enriched with microbes found on human skin. Jacomo was colonized this way as he lay on his mother's breast in the delivery suite, wondering what the hell had just happened. But it's becoming clearer that babies born by Caesarean have a different immune development from those born by vaginal delivery, with a higher likelihood of obesity, allergy, atopy and asthma, and lower diversity of the intestinal gut microbiome.[15]

The official name for the initial colonization of a newly born baby is 'dispersal'. The precise order and type of microbes that disperse themselves throughout the gut depend on the maternal and environmental exposure. So where you deliver and nurse your baby will have an impact on the newborn microbiome. Sixty Slovenian babies born at three maternal units had their microbiome serial sequenced during their first month of life. The study's researchers observed that during the first month of life the relative proportion of *Enterococci* was closely associated with the locality of the hospital the babies were born in.[16] The dispersal of the microbiome is determined in part by the environment of birth, and babies born with a home labour will have an altogether different microbiome from those delivered on a labour ward. However, the long-term implications of this finding are not known.

Premature babies are cared for in neonatal intensive care units. The constant de-sterilization and repetitive hand-washing that characterize the ICU make it an extremely austere environment for the microbiome. Despite this, the incubator, the ward and the hands of hospital staff still provide sources of gut-colonizing bacteria for the baby. In particular, *Candida* and *Saccharomyces* species are abundant, with an undefined impact on early microbial development. 'Micronates', or very underweight premature babies, have important anatomical and physiological challenges compared to babies born at full term. Thanks to the extraordinary work of neonatology teams across the world, many very premature babies survive and lead happy and normal lives. But babies born prematurely have an increased risk for type II diabetes, cardiovascular and cerebrovascular diseases, hypertension, chronic kidney disease, asthma and pulmonary-function abnormalities, neurocognitive and psychosocial disorders and poorer social adaptation. It's tempting to speculate on the role of the microbiome in the causation of these conditions. If it does play a part, it is far more likely that this is because of a profound disruption to metabolic and immune programming rather than early-life skin contaminants in intensive care.

Babies born without any microbes, however, would not be normal. Surprisingly, there is a way to re-create this extraordinary circumstance within the laboratory. 'Xenobiosis' is the study of germ-free animals. These experimental creatures have been artificially bred, and live their entire lives in a plastic bubble where their air is filtered, their bedding, food and water are irradiated and everything they touch is sterilized. The only circumstances under which they would meet a colonized animal would be for an experimental purpose. They're continuously screened for infections and treated with a cocktail of antibiotics. Their young are delivered by Caesarean section into a sterile world of isolation, so their gestated pups aren't subject to maternal microbiome interference. Many different types of animals – from pigs to mice and rats – have been subjected to this Kubrickesque existence.

The method was initially conceptualized by Louis Pasteur in 1885, although he didn't believe it would be possible for animals to exist without bacteria – and he was partly right. Mice born into this world of isolation look externally like mice. Their genome can programme the growth of all their key organs, their lifespan is equivalent to that of a colonized animal and, on the bright side, they are very unlikely to get cancer. However, they are far from normal. Their growth and development are stunted, and they have major anatomical and physiological differences. For example, the caecum – the first part of the colon or large bowel – is enlarged by four- to eightfold in germ-free mice, and their small intestines have a considerably smaller surface area than those of colonized animals. Germ-free mice are thinner, because the hormones that signal appetite in the brain are impaired and they are less efficient at extracting energy from their diet. They need to be sustained with essential vitamins, like vitamin K, to make sure that their blood clots and they don't bleed to death. They have severely altered immune systems and any infection, no matter how mild, can be catastrophic.

Germ-free animals also behave differently and, most spectacularly, this can be reversed by colonizing their gut with normal bacteria.[17] Remember, these mice are bred to be genetically identical, and experiments have demonstrated that microbiota are necessary for normal stress responses, anxiety-like behaviours, sociability and cognition – in other words, behavioural factors that influence *risk-taking*, among other behaviours. Humans, of course, are not germ-free, and we shouldn't forget the extreme nature of these models when drawing inferences about human health. But the important takeaway is that the developing gut microbiome might have vitally important implications for the mental health of our young.

Magical milk

Breast is best. But breastfeeding is also politicized, moralized, sometimes painful or even impossible. Infant formula helps women (and babies) who can't breastfeed and it saves lives. But we have become remarkably dependent on it: 2.7 billion tonnes of formula milk is manufactured each year, with a global market value of more than US $100 billion.[18] Historically, commercial bias in the marketing of milk formulas has been a problem. So much so that in 1981 the UN World Health Assembly had to recommend an international code of conduct to govern the promotion and sale of breast-milk substitutes. Despite this, aggressive marketing is still employed today, and in February 2022 it caused a national disaster in the US. A safety scandal at Abbott Laboratories in Michigan forced the company to shut its formula-feed factory. The problem was that Abbott supplied 15 per cent of the entire US formula-feed market, and when this was combined with COVID-19, the war in Ukraine and panic buying, the entire supply chain collapsed. The country experienced a nationwide shortage that was so catastrophic that New York City declared a state of emergency, and the military began flying in supplies from Europe as part of 'Operation Fly Formula'.

Scares like these, along with large doses of fear-mongering, lead many women to worry about doing the right thing, and often the question of breastfeeding becomes an unnecessary source of guilt and anguish. However, breastfeeding is an affordable, lifesaving intervention in deprived communities, particularly where clean water is sparse, and it irrefutably plays an important role in infant health and survival in these circles. Some of its benefits – prevention of infection in the baby, and of sudden infant death syndrome – are well established across all social classes and economies. There are some shared benefits for both mother and baby, such as a lower risk of obesity and cardiometabolic disease. And in the mother, breastfeeding reduces the risk of ovarian and breast cancer and osteoporosis. The longer you're able to breastfeed, the greater the benefit, but also the longer the baby's teeth – and everyone has a limit.

Breastfeeding is so beneficial because it's the gut microbiome equivalent of the Big Bang. The diversity of microbes in the newborn gut drops until breastfeeding takes hold. Breastfeeding triggers such a massive microbial bloom that, by week one, the baby has a gut microbiome that will remain constant until around one month of age. We don't yet have the full picture of how this happens. We know some bacteria are transferred directly from the skin of the breast. In populations around the world there's a strong correlation between the bacteria found on the breast and those identified in the faeces of newborns. There's a powerful link between the make-up of an infant's gut microbiota and the extent and duration of breastfeeding. For example, a study of babies in Thailand found that the abundance of *Bifidobacteria* was closely associated with how long babies had been breastfed for, and that babies fed with infant formula had significantly lower abundances of this species in the gut.

But we also know there has to be another factor – which has more to do with the molecular content of milk – driving the growth of bacteria within the gut of newborn babies. A Bangladeshi birth cohort, sampled every month for sixty months, identified a network of fifteen covarying bacterial taxa, which means that there is more than just one species benefiting from breast milk, and nourishing such a diverse set of bacteria requires a complex nutritional matrix, as found in breast milk.[19] Similar studies from Malawi, Thailand, Slovenia, China, the USA, Finland and the UK have also shown that common microbial life forms coexist in the developing gut, implying that breast milk is able to repeatably influence the growth of beneficial microbes, no matter where in the world the baby is being fed.

The exact content of breast milk is still being defined. Part of the challenge is that breast milk is individualized to the mother, and its composition changes over her lactation period. However, we can think of it as a 'live tissue' that contains genetic material from the mother as well as a microbiome, which is, more literally, alive. Infant formulas are good and getting better, but they can't yet precisely replicate the complexity of breast milk or its live component. The aqueous content of breast milk is made up of lactose, fats (lipids and steroid hormones) and a group of complex sugars known as human milk oligosaccharides (HMOs). The composition of breast milk helps to grow specialist bacteria that metabolize HMOs and species adapted to living within the mucus of the gut. These bacteria have many different enzymes that can break down carbon sources – which become abundant once a baby starts weaning and eating other food. Breast milk also shapes the bacterial community because it contains lots of antimicrobial factors – like lysozyme, lactoferrin and antibodies – that stop pathogens from growing or symbiotic bacteria from becoming too powerful.

Milk also contains many types of antibodies, the protective proteins produced by the immune system. The maternal microbiome orchestra is extremely clever, and it has a long memory for pathogens, which it shares with its newborn through milk. Mice that have been exposed to nasty pathogens – such as the aggressive bacterium enterotoxigenic *Escherichia coli*, which causes terrible diarrhoea – will develop antibodies through their immune system. Maternal mice can share these antibodies with their pups through their breast milk, so that suckling pups are able to mount a defence against this bacterium if they come into contact with it in the future. In other words, mothers can share their immunological memories of pathogens to protect their naïve children through their breast milk.[20]

While the impact on the breast-milk microbiome is minor, there are significant differences in the way that women breastfeed around the world, and this might yet be found to have an impact on how specific strains are transferred. Bottle-feeding also influences the microbiome, as live bacteria can hitch a ride on the plastic.[21]

If you're the father or non-gestating parent of a baby and you are feeling left out while you read this, don't worry. Your microbiome is important too, and it will also greatly influence the development of your baby's health. When you hold her hand for the first time, the microbes you share will go straight into her mouth when she sucks that hand. Tactile interaction plays an important role in dispersing and selecting microbes. In fact, as he or she grows into a toddler, furry pets and day-care attendance all play a role in microbiota assembly and in ensuring the diversity of the gut.

The first 100 days

Colonizing our young with symbiotic bacteria as quickly as possible is critical to our health. So critical, in fact, that in the animal kingdom mothers will take things directly into their own paws, particularly if the required microbes can't be transferred in breast milk: maternal panda bears feed their offspring their faeces, to pass them bacteria that help them metabolize cellulose, the molecule that provides the nutritional value from their main foodstuff, bamboo. Without the bacteria, panda babies wouldn't be able to digest their most important foodstuff. Hippos, elephants, dogs, koalas and insects all do it too, so this scatological horror show must have some benefit. At least we humans can all breathe a sigh of relief that we have evolved out of this . . . except that we haven't.

After the initial period of microbial dispersion triggered by birth, breastfeeding and nursing, gut microbial selection begins. This is survival of the fittest, where different taxa do battle in a bid to take ownership of gastrointestinal fiefdoms. During this early stage, antibiotic use or illness has a significant impact on taxa that exist in low abundances, and it can decimate whole ecosystems, which, in some cases, the gut will not recover. Outside illness, the two dominant forms of selection are the developing immune system and the diet. As we're seeing, the relationship that the microbiome has with both of these factors is intricate and dynamic.

After selection, an infant's gut ecology 'drifts' as it's exposed to more random and less sustained ecological events that influence its diversity. Finally a period of 'diversification' happens, as microbes mutate and share genes that let them compete and survive. This happens with the arrival of teeth and the change to solid foods, which forces the colonic microbiome to diversify as it adapts to metabolize plant foods and new substrates. These solids change the conditions in the colon, selecting for bacterial populations with relevant metabolic activities, and the microbial diversity of the intestine increases steadily. The exact length and timing of these critical events in our gut development are variable and overlapping, but by around three years old a toddler will have a gut microbiome comparable to that of an adult in its ecology and structure. From this age, the microbiome is far more stable and less prone to influence from individual environmental stresses. Fundamental damage to the microbiome during the first three years of development can have long-lasting implications for the health of those children.

Despite my barely adequate parenting and constant failures, my kids continue to thrive – and it's a miracle to me! Their risk of chronic disease might have been influenced by their route of delivery into this world, as it might have been affected by the way they've been fed and nurtured. I can only tell you that the Girl and I would not change most of the decisions we made for our children and, like most people, we didn't have much choice anyway.

The twentieth-century approach to protecting our newborn babies has been to pursue the advertising slogan of a famous brand of household bleach, which is to 'kill all known germs dead'. This strategy now requires some subtle readjustment, because our young are not germ-free laboratory animals – this is as absurd as it is impossible; you can't sterilize a human being unless you boil them in a bag. Perturbations in the maternal microbiome and the infant-gut microbiome in the first 100 days of life may well define the immunological, neurological and metabolic capacity of the future adult and, in turn, their risk of developing a large number of chronic diseases. Our children will carry the consequences of this strategy for longer than any other generation that has ever walked the Earth.

If we're to 'save' the microbiome, the most important time to do so is in early life and most probably within the first 100 days. Improving the health of the neonatal microbiome isn't solely a matter of increasing breastfeeding rates or smearing children born by

Caesarean section with vaginal microbes. An even better strategy would be to prioritize the maternal microbiome, which is critical to the development of a healthy human. The microbiome needs to become a fundamental component of neonatal health, and it should be considered from the point of conception. This means that the global corporations that produce the foods, the medicines and the extreme environmental conditions that deleteriously shape our children's microbiomes should be held accountable. This has to be done urgently, because our children are inheriting a microbial orchestra that is playing a different tune from the one that serenaded their grandmothers and mothers before them – and, soon, the music will be lost for ever.

7. Symbiotic Sentience

With a trembling hand I took the drill from the neurosurgeon.

'Just don't fuck it up,' he said.

Sweating, I pressed the drill bit against the skull, right above the temple where the hair had been shaved and the skin had been peeled back. Luckily for the patient – a twenty-three-year-old woman we'll call Betty – and for me, modern orthopaedic drills used for cutting into skulls are extremely clever, and a clutch safely cuts the power to the drill as soon as it reaches softer tissues that lie beneath the skull.

The objective of the operation was to release a large volume of gelatinous, clotted blood that was pressing on the brain. Betty had sustained a head injury after being knocked from her motorbike, tearing blood vessels in the lining of the brain. There was a finite amount of time before permanent brain damage would be caused by the pressure of the expanding clot, which would ultimately force her brainstem through the base of her skull. As I stood under the operating lights, overheating in my gown, I thought about how bland the brain appeared. A lifetime of experience, love and learning lost in grey-and-red blancmange.

The only way we've been able to map the dense network of eighty-six billion neurones that make up the original biological supercomputer is, ironically, by using artificial intelligence. But even the insights into the brain's 'interactome' aren't enough to completely break the software code for the brain's operating system. The collision of neuroscience and the microbiome is, however, providing startling new insights that are redefining our understanding of how the brain develops, how it degenerates, how it thinks and even how it responds to injury. This has led to an existential question: if our brain is in conversation with our gut microbiome, what role do microbes play in human sentience? Could we be experiencing a collective symbiotic consciousness? To answer this, let's start by considering how our nervous system works and how it's connected to the gut.

The blood–brain barrier

The brain is physically separated from the rest of the body by a microscopic anatomical and physiological wall. Blood vessels and capillaries that carry oxygen and fuel to the nerve cells of the brain are guarded by tight junctions designed to prevent unwanted invaders from getting in, because pathogens that infect the brain are a threat to life. Most commonly it's viruses and bacteria that cause the damage, but not always. In 2012, doctors from Sichuan University in China reported a case of a fifteen-year-old girl who had, for a year, been complaining of numbness in the right side of her face and upper limbs.[1] She then started having seizures, which led to an MRI that revealed an irregular lesion on the left side of her brain. On further questioning, the young woman confessed to eating inadequately cooked frogs, from which she contracted a tapeworm. The larvae had travelled to her brain and there they grew up to 12 cm in length, ultimately requiring surgical removal. This condition is called neurocysticercosis and is more commonly caused by ingesting a pork tapeworm (*Taenia solium*), but the lesson here has to be: don't eat raw frogs.

Fortunately, stories like these are rare because, as a general rule, the brain is extremely good at keeping out pathogens, all the while bathing itself in a constant supply of the finest metabolites known to humanity. It's so good at its job that until recently it was assumed that the microbiome couldn't penetrate the blood–brain barrier. The gut microbiome, however, has co-evolved with the brain over millions of years, which means it has a multitude of tools at its disposal for hacking this security feature.

How the brain and gut talk to each other

The peripheral nervous system describes the nerves outside the brain and spinal cord (known as the central nervous system). Its job is to regulate the physiological processes that happen in our bodies involuntarily which keep us alive – such as our heart rate, blood pressure, breathing and digestion – and it is also responsible for sexual arousal. It's made up of three parts, called the *sympathetic*, *parasympathetic* and *enteric* nervous systems.

The sympathetic nervous system is responsible for initiating our 'fight or flight' response; in the wake of her motorbike accident, Betty's sympathetic nervous system was literally fighting for her life.

Meanwhile the parasympathetic nervous system manages the 'rest and digest' reflexes; it's dominated by the vagus nerve, which controls 70 per cent of the parasympathetic nervous system and manages our heart rate, our digestive system and our mood. It even has an immune function. The vagus nerve takes a long and tortuous course through the body. While part of its job is to send signals 'down' the network, almost 90 per cent of all fibres are used to send signals 'up' to the brain from various parts of the body for processing.

The enteric nervous system (ENS) is embedded in the wall of the bowel and is sometimes referred to as our 'second brain'. It's made up of more than 100 million neurones and contains more nerves than the sum of all other peripheral nerves in the human body combined. Because it's responsible for managing so much of the gut's operating system, there's a huge diversity of nerves within it – twenty different subtypes of nerve use more than thirty different types of neurotransmitters – and it is connected to both the parasympathetic and sympathetic systems. These in turn work against each other and, by doing so, create an equilibrium. Every day our body's nervous system performs a balancing act with one foot on the accelerator (sympathetic) and one foot on the brake (parasympathetic). In this analogy, the microbiome in the gut serves as the clutch, regulating how much power is transferred from the engine to the wheels.

We're now starting to understand quite how deep some of the wiring from the brain to the gut runs. The vagal and spinal nerves that carry signals from the brain reach all the way to the lining of the intestine, called the mucosa. These nerves directly control the gut's secretions and absorption of food and fluid, and even blood flow to the gut, which in turn influences the type and number of microbiota living in our bowels. For example, pain receptors in the gut can sense the bacterial pathogen *Salmonella*, which is very good at picking the lock of important immune cells that live there, called M cells. The enteric nervous system works like a burglar alarm: when it detects the *Salmonella* bacteria it fires a message to the brain to warn it that there's trouble.[2] But it also reduces the number of M cells to stop more bacteria from getting in, and closes the gate to the castle. Even more cleverly, it promotes the growth of defensive filamentous bacteria that line and guard the gut against the invaders. So the enteric nervous system also has immunological functions and is a critical component of our defence against pathogens.

Your brain relies on far more than just five senses to understand the world. Because it is so intimately connected to the microbiome, it has trillions of sensing microbial organisms that monitor and interact with your environment. But the connection goes even deeper: because the gut microbiome plays a part in regulating all three components of the nervous system, it also has a say in your flight-or-fight response, your rest-and-digest response and the day-to-day function of your immune system. This means that the microbiome influences how you experience stress, and how tired, horny or angry you are. Cracking this microbial code is becoming increasingly urgent, as it's this cross-talk that also defines our risk of both acute and chronic neurological disease.

Brain attack

Each year thirteen million people will have a stroke – or brain attack – and around 5.5 million will die from this event.[3] A stroke occurs when the blood supply to part of the brain is interrupted, leading to the death of irreplaceable neurones within it. We've recently discovered that clots retrieved from arteries in the brains of patients during a stroke have a diverse microbiome of their own.[4] This is peculiar, because they should in theory be completely sterile, and no one is quite sure how these microbes got there or what they are doing.

Stroke usually occurs in older people who have an ageing microbiome and less resilience to any injury in general, and multiple studies suggest that the severity of a stroke is greater in people who have a less diverse gut microbiome. For example, patients with stroke have increased abundances of *Streptococcus*, *Lactobacillus* and *Escherichia* and lower abundances of *Eubacterium* and *Roseburia* taxa in their gut, and these changes also correlate with the severity of a stroke. Many of these associations relate to how our lifestyle and our diet modify stroke risk. For example, fibre in our diet protects against stroke risk because it is metabolized by bacteria such as the *Eubacterium* and *Roseburia* taxa into anti-inflammatory molecules that protect the brain.[5] If you are not eating enough fibre, the relative abundances of these bacteria fall and your stroke risk rises.

A brain attack also causes reciprocal changes in the gut microbiome, because when the brain suffers an injury, blood flow to the gut is reduced, leading to the release of unstable atoms that damage cells, known as free radicals.[6] This also allows bacteria that like a low-oxygen state to bloom in the gut, changing its diversity and in turn the systemic levels of inflammatory molecules. Stroke damages the gut barrier and, for this reason, many bacteria that cause chest infections in stroke patients originate in the gut.[7] It also means that the gut microbiome disrupts the integrity of the blood–brain barrier and alters the severity of the brain injury.[8] So it follows that, in mouse models at least, antibiotics reduce the severity of stroke and, amazingly, this neuroprotective effect can be transferred through faecal transplantation.[9]

Because of this close link between the gut and the brain, scientists are now starting to identify biomarkers that measure the activity of the microbiome to predict stroke severity. For example, the gut microbiome breaks down animal proteins and nutrients (such as choline) in our diet, to produce inflammatory metabolites such as Trimethylamine N-oxide. High levels of this compound in the blood are associated with increased stroke risk. Maybe it's time to look deeper at how the gut microbiome communicates with the brain – and how it influences brain development.

The gut–brain axis

Our gut, microbes and brain have three different ways of talking to each other. They can do so through the manufacture of neurotransmitters (like serotonin); through the hormones in our endocrine system; and through the immune system. The microbiome can work across all three of these at the same time, and we are now slowly unravelling some of the detail of how this works.

Neurotransmitters produced by the microbiome don't cross the blood–brain barrier under conditions of health, but they can act directly on the enteric and peripheral nervous system, which in turn sends signals back to the brain. Microbes can also create metabolic precursors of neurotransmitters that can cross the blood–brain barrier, and the amino acid tryptophan is increasingly recognized as a 'master switch' for this process because it can be made into multiple different types of neurotransmitters.[10] The vagus nerve hooks the brain into a hormonal 'endocrine' network that makes steroid hormones that influence our body's stress response. However, the gut also contains its own endocrine cells that produce hormones and bacteria that live along the lining of the gut and are able to interfere with this system, influencing everything from our appetite to our mood. Very sub-specialized 'neuropod' cells plug these endocrine cells directly into the enteric nervous system, which means that bacteria can communicate directly with the nervous system.[11] The immune system influences the function of all organs in our body, and our nervous system is no different. Microglial cells, for example, are the resident cells of the innate immune system, and these make up to 15 per cent of all cells found in the brain. Microbes are able to interact directly and indirectly with these cells to modify brain health and our behaviours. Collectively, all these pathways can be manipulated by the microbiome to influence our state of mind.

The psychobiome

When we're ill we tend to display what doctors call 'sickness behaviour': a suite of systemic responses, including social withdrawal and loss of appetite. There's an evolutionary benefit to this, because isolation prevents us from spreading infection. Losing our appetite when we're ill might also promote metabolic changes in our bodies that keep us healthy – for example, by limiting the nutrients available for pathogens that wish us harm. But some bacteria have learned to subvert this process for their own benefit, and they do it through the vagus nerve. *Salmonella enterica* is responsible for a quarter of all cases of diarrhoea, causing more than ninety-four million gastrointestinal infections each year. In mouse models, *Salmonella* blocks the attempts made by the vagus nerve to detect inflammation in the brain and prevents normal sickness responses from developing. More specifically, it inhibits the loss of appetite response – so the mouse carries on eating as it would if it were healthy.[12] In other words, these bacteria alter our higher brain function for their own survival benefit.

We don't need to be ill for the microbiome to influence our moods, and more complex behaviours, via these same pathways. Across Europe, it's been noted that people suffering from depression have distinct changes in their gut bacteria.[13] After correcting for the confounding effects of antidepressants, an analysis of the microbiomes of 2,124 people determined that levels of *Coprococcus* and *Dialister* spp. – bacteria that produce the pleasure chemical dopamine – were depleted in depression. At the same time, higher abundances of bacteria such as *Faecalibacterium* and *Coprococcus* spp. were consistently associated with higher quality-of-life indicators.

Because dopamine controls our most important reward pathways in the brain, food and drug studies have started to explore the connection between the microbiome and our vulnerability to addiction. For example, mice given prolonged courses of antibiotics show a higher susceptibility to cocaine addiction.[14] It's unlikely, however, that it was the antibiotics that made Tony Montana quite so psychotic.

We've made similar observations about the microbiome's role in regulating our anxiety. Bacteria such as *Bacteroides ovatus* produce metabolites that break down dietary sources of tyrosine – which is found in foods such as soy, chicken, avocados, bananas and milk products. This metabolite interferes with a cell responsible for wrapping the axons of nerves in a type of fat. The fat allows the nerve cell to send its electrical signal along its arm-like axons, which connect with other nerve cells via a synapse to send signals.[15] Mice that don't have enough *B. ovatus* are unable to lay down this fat effectively, and this prevents their brain from developing normally, producing anxiety-like behaviours.

People with inflammatory bowel disease and other chronic gut diseases have higher rates of mental-health problems like depression and anxiety than the general population. Irritable bowel sufferers also experience this, and commonly endure high rates of hypervigilance, pain anticipation or emotional hypersensitivity. And there's increasing evidence that says the gut–brain axis plays an important part in causing these symptoms. For now, though, it's not clear which comes first: the gut or the brain. My view is that they should not be considered separately and that these things occur in parallel.

The gut microbiome grows a brain

As a dyslexic person, I have personal experience of the obstacles that many neurodiverse people face in getting through school or holding down a job. My head constantly fizzes with unconventional ideas (like the microbiome), and in my academic career I have had to retrain my brain to perform the simple tasks that most people take for granted, like spelling. In time, I have come to realize that my dyslexia is a superpower, and now I would not choose to be without it.

In 1999, the sociologist Judy Singer coined the term 'neurodiversity' to de-stigmatize neurodevelopmental conditions such as dyslexia, Attention Deficit Hyperactivity Disorder (ADHD) and Autism Spectrum Disorders (ASD). Until that point, these conditions were typically labelled disabilities or pathologies; Singer's goal was to celebrate difference and acknowledge that as well as presenting challenges, these conditions create extraordinary potential. A huge amount of work is now being performed to explain how these conditions occur, and scientists are asking whether the microbiome contributes to their development.

New nerves – created during foetal development – slowly migrate into position as the baby gestates. These must also make new chemical connections between each other, called synapses, and it's this pattern of connections that creates memories, thoughts and our higher brain functions. Synaptic pruning happens when connections that are incorrect or no longer useful are destroyed and it is an important method for regulating this process. One hypothesis is that the development of our neurones is co-dependent on the gut microbiome, which undergoes parallel changes in its diversity and structure with the developing brain. Or to put it another way, the microbiome modifies the structure and function of emerging neural circuits, creating neurodiversity – and my dyslexia. The race is now on to define the precise mechanism through which this occurs.

By the age of two the brain is 75 per cent of its adult size. At ten years, a child's brain represents 5–10 per cent of body mass, and it consumes twice the glucose and one and a half times the oxygen per gram of tissue compared to an adult's brain. This is hungry work, and it accounts for up to 50 per cent of the body's metabolic output. The gut produces much of the energy for this massive and rapid brain growth, and it's hard to sustain a healthy growing brain with the wrong gut bugs.

So the timing and the quality of the assembly of the microbiome are important. As previously described, the altered behaviour of germ-free mice can be reversed by colonization with normal bacteria into their gut. However, this intervention is age-dependent. Post-weaning recolonization, for example, is more effective at restoring anxiety-related behavioural deficits than recolonization later in life.[16] These behaviours are particularly influenced by the neurotransmitter serotonin, whose functions cannot be restored by recolonization of the gut after weaning. The implication is that the greatest time for microbial influence over our behaviour and psychological well-being occurs in the earliest parts of our lives and, when this is over, it may be over for good.

For now, the role of the microbiome in brain development remains contentious, and the greatest controversy is saved for ASD. This refers to a condition manifested by a range of symptoms characterized by some degree of impaired social behaviour, communication and language, and by a narrow range of interests and activities that are both unique to the individual and carried out repetitively. There is a huge amount of work across multiple scientific fields attempting to define why children develop autism, and the microbiome is merely one avenue of investigation. Regardless of the cause, the rates of ASD in the developed world are climbing. It now affects around one in fifty-seven children (1.76 per cent) in the UK, with the highest rates in pupils of Black ethnicity (2.1 per cent) and the lowest in Roma/Irish Travellers (0.85 per cent).[17] The significant differences in autism prevalence across ethnic groups and geographical location implies that the exposome – and, by definition, the microbiome – is likely to have an important role.

The mean age of diagnosis of ASD is five years old,[18] just after the microbiome has finished assembling, and multiple studies have found differences in the faecal microbiome compared to children without ASD. In 1943, when the Austrian-American psychiatrist and physician Leo Kanner published the first description of eleven children with autism,[19] he reported on a thirty-three-page letter from the father of a child called Donald T. In this extraordinarily detailed and moving description of his son, the father noted: 'he has never shown a normal appetite'. It's now recognized that many children with ASD also suffer from gastrointestinal symptoms ranging from constipation to diarrhoea, and that these might be under-reported.[20] Gut dysfunction and autism appear to be linked, but like all things gut–brain, it's not clear which comes first.

A Chinese study of more than 773 children with ASD and 429

neurotypical children suggests that this might be more than mere coincidence. The researchers studied the dynamics of the gut microbiota across different ages,[21] and found that the gut microbiome of children with ASD evolves in a way that is indeed influenced by diet and gut function. But this evolution is strikingly different from that of neurotypical children. Children with ASD have an immature gut microbiome, with less diversity, and the most important changes take place before the age of three. These observations on the microbiome closely associate with the severity of behaviour change, sleep patterns and altered bowel function in the ASD group. Researchers were able to identify the children with ASD through the detection of the bacteria *Veillonella* and Enterobacteriaceae, along with just seventeen microbial co-metabolites found in their stool, which differentiated them from the neurotypical children.

The severity of ASD also appeared to be significantly linked to the function of the microbiome and bacterial pathways for tryptophan metabolism, which is relevant because it is used to make neurotransmitters. There's more evidence that other metabolites created by gut bacteria influence brain development. For example, bacteria in the gut putrefy proteins and one of the by-products of this process is a metabolite called p-Cresol. When this is given to mice, it selectively causes ASD behavioural symptoms and social behavioural deficits.[22]

However, it also changes the composition of the gut microbiome, and when faeces from p-Cresol-fed mice is transplanted into control mice, they not only produce more p-Cresol, but also develop the same behavioural symptoms. Most interestingly of all, when faeces from healthy control mice is transplanted back into p-Cresol-fed mice, their behaviour changes back to that of healthy mice.

So it should not be a surprise that transplanting faeces and the microbiome from human donors with ASD into germ-free mice also prompts autistic behaviours in the mice.[23] This is likely to be an 'orchestral' effect.

These clever experiments don't, though, explain the role of the parental microbiome in the development of conditions of neurodiversity. This needs to be understood, because it appears that a misdirected maternal immune system is important in the development of ASD. In one Swedish analysis of more than 1.8 million people, children of mothers who were hospitalized for any infection during pregnancy had a 79 per cent higher risk of being diagnosed with ASD.[24] Intriguingly, this observation can be replicated in mouse models, and certain types of bacteria known as 'segmented filamentous bacteria' seem to be particularly important in changing the behaviour of the immune system in pregnant mice. These bacteria take the brake off the adaptive immune systems and induce the production of a particular type of lymphocyte (T helper 17 cells), causing abnormal behaviours in their offspring.[25] Interestingly, these bacteria do not behave the same way in non-pregnant mice, but it is not clear precisely why.

We don't know the exact source of the infection of those women included in the Swedish study, but where there are pathobionts, there is inflammation, and this is one of the major methods through which the microbiome influences behaviour. The problem is that there could have been too many pathogens driving inflammation or not enough symbionts preventing it. It is also possible there is a confounding factor (such as a medicine) or something in the way these infections were treated that explains this observation. We don't know yet if COVID-19 is going to serve as a driver of maternal inflammation, but it's possible. If you're a woman thinking about having a baby, it's another very good reason to get vaccinated.

As we come to understand that the microbes in our gut are an important cog in the ASD and neurodiversity engine, we're also faced with an important question: what do we want to do about it? Just as diversity determines the health of our environmental and internal ecosystems, so neurodiversity is an important measure of the resilience of our society, our schools and our workplaces. Despite humans' genetic similarity, the brain exhibits a wonderful variance: we don't all think the same, and our species relies on this to survive. As we start to unravel the mysteries of how the microbiome influences our neurocognitive development, we can hold on to the fact that it offers a new therapeutic avenue where few exist. This is incredibly important for those people and their families who have been adversely affected by conditions of our developing brain, which can certainly be disabling, as Donald T's father so eloquently described.

Psychobiotics

The emerging data on the microbiome raise the intriguing possibility that we can now selectively engineer specific strains of bacteria to deliver the healthy metabolites we need in order to improve our higher brain functions. To date, trials in humans of probiotics, which are sometimes referred to as psychobiotics, are lacking, and the evidence base for their wider use comes largely from animal studies. However, these are rather extraordinary pieces of research.

Both *Bacteroides fragilis* and *Lactobacillus reuteri* have been observed to reverse some of the behavioural and gastrointestinal changes reported in animal models of Autism Spectrum Disorder, but these bacteria work in discretely different ways. For example, in a mouse model of ASD, *L. reuteri* rescues social behavioural deficits by working through the vagus nerve to increase the amount of a pro-social hormone called oxytocin in the reward pathways of the brain. It also changes behaviour by creating a metabolite called tetrahydrobiopterin, which is a co-factor involved in the production of several neurotransmitters such as dopamine, serotonin and nitric oxide. It is also true that psychobiotics don't simply work on any ASD-associated behaviours. For instance, they did not have any impact on hyperactivity behaviours because these are almost exclusively controlled by the host genetics of the animal.

To date, many of the studies in this very early field have focused on the ability of probiotic strains to produce neurotransmitters. For example, the probiotic *Bacteroides uniformis* CECT 7771 cured binge-eating in rats and reduced their anxiety-like behaviour, at least in part, because of changes in the circulating levels of dopamine, serotonin and noradrenaline produced by the bacteria.[26]

It is important to add that we don't have any evidence that these probiotic strategies are effective in humans. The best way to improve our brain health may be to prevent conditions from forming in the first place. And the best way to do this may simply be to get a good night's sleep.

Sleep tight and don't let the bedbugs bite

Our bacteria and their functions are closely associated with our sleep cycle, and our gut microbiome's activity subtly fluctuates over a twenty-four-hour period. The brain controls circadian rhythm by means of a neurological 'master clock' called the suprachiasmatic nucleus. However, our sleep cycles are also influenced by diurnal variations in the function of the microbiome and daily oscillations in its metabolite production and epigenetic programming of the gut. Similar circadian rhythms exist in fungi and cyanobacteria. If this microbiome 'rhythmicity' is interrupted, it perturbs the regular work of organs like the liver, and it causes low levels of inflammation in the brain.[27] This goes some way to explaining why those who can't sleep are at greater risk of metabolic syndrome and heart disease.

If, like me, you're a shift-worker and you work nights, then your microbiome is also being influenced by your work pattern. Scientists can tell if you're a day-worker or night-worker just by looking at your microbiome: an analysis of male security guards found that the *Faecalibacterium* genus abundance in the gut is a biomarker of day-shift work, and *Dorea longicatena* and *D. formicigenerans* were significantly more abundant in individuals working the night-shift.[28] *F. prausnitzii* is one of the most abundant and important commensal bacteria of the human gut microbiota and is strongly associated with gut health.

So how do bacteria know what time of day it is? Some bacteria, such as *Bacteroides ovatus*, possess their own internal clocks that do not rely on light. Instead their populations 'oscillate', based on environmental cues that come from our diet and lifestyles. Diets that are generally thought to be healthy, such as plant-based low-fat diets, promote their growth, while unhealthy low-fibre, high-sugar, high-fat diets inhibit their growth. These bacteria influence our own circadian rhythm or our metabolism, and this is another reason why eating junk food when you are working nights will not help you sleep.

In fact if you want to repair your nerves or your brain, eating less might also help: we've recently learned that intermittent fasting promotes nerve regrowth. This process relies on a metabolite called indole-3-propionic acid (IPA),[29] which is produced by a specific bacteria called *Clostridium sporogenes*. In a mouse model, IPA administered after an injury to the sciatic nerve significantly sped up the recovery of sensory function because IPA helps coordinate the immune-dependent regeneration of the axon. It is amazing to me that the gut microbiome may yet be a target for devastating nerve injuries. But let's look now at what you do if your injuries are in the mind.

Puppets on microbial strings

If the microbiota within us are able to influence our feelings of satiety, anxiety, sexual arousal and pleasure, we may well start to question how much free choice any of us really have. What if we're simply serving our microscopic overlords? Nature shows us several horrifying precedents for this behaviour. Take the *Ophiocordyceps unilateralis* fungus, for example, which infects ants in tropical forests. Through a combination of brute force and enzymes, its spores eat their way through the ant's exoskeleton. Once on board, the fungus enters the ant's brain, where it takes control of the ant's higher brain functions by releasing neurotransmitters and metabolites.[30] In doing so, it creates a 'zombie ant', which it then directs to leave the safety of its colony. Sometimes it will induce a seizure to encourage it to fall, but it will also alter the ant's climbing behaviour to send it marching down to the cool, damp forest floor, which is better suited for the fungus. Once there, the ant is instructed to dig its jaws into the vein of a large leaf and it dies. The fungus then feasts and slowly grows out of the ant's skull in a long spear-like stroma, before dispersing its young into the forest, where they will repeat the cycle. The ant's body is left looking as if it has been killed by a javelin to the back of the head.

This gruesome host-manipulation could be taking place in you and me right now and it's feasible that we're all just existing in a microbial matrix. After all, there is theoretical evolutionary benefit for microbes that are able to reproduce and disperse by exerting their control over our behaviours. You'll be relieved to hear, however, that this is very unlikely. It's far more probable that microbes influence our behaviour through a process of symbiosis. For instance, our gut symbionts are selected to be there by our genome and our exposome, and in turn they are naturally selected to influence the local gut environment. The impact they have on our behaviour is a side-effect of this relationship; for example, it changes how we regulate the environment of the gut. Similarly, if a person's behaviour changes when they're missing a particular type of bacteria, that's more likely to be because the physiology of the gut has changed than because that individual has had their mind changed. In other words, we exist in a finely balanced state of evolved co-dependence with our microbes.

Unfortunately, this careful balance has been dramatically disrupted by the discovery of antiseptics, antibiotics and urban living. And this is exacerbating the explosion of mental-health conditions that we're seeing among young people. The frail human symbiont brain has been unable to adapt to these insults to our internal ecology, causing a fundamental disruption to its development and a pandemic of neurological and psychological disease.

It's hard to speak in more concrete terms because we're overwhelmed by a confounding range of causative factors. Digital addictions are distorting our dopamine sensitivity and serotonin-feedback loops; while relationships, trauma, social isolation, education, socioeconomics, drugs, antibiotics, alcohol and culture all modify the mental-health exposome. The psychobiome is doing its best to survive in the exposome we have created, but it isn't happy: between 2017 and 2018, 7.3 million adults in England – 17 per cent of the entire adult population – took an antidepressant.[31] And this has only climbed since, given that just under one in three adults reported a clinically significant level of psychological stress during the pandemic lockdowns of 2020 and 2021.[32]

Unless, as a society, we choose to invest in social-care systems to prevent, detect and manage mental-health problems or cognitive decline, we will continue to fail the most vulnerable members of our society. These systems must now also prioritize the health of the gut as an enabler of a healthy mind. This solution doesn't require expensive microbial biotech, but it needs to make use of the knowledge we already have when it comes to optimizing the maternal microbiome and gut–brain development in early life and preventing the neurological diseases of old age.

I want to tell you that Betty did well after the operation, but her injuries were too great, and she died in the intensive care unit forty-eight hours after her surgery. The treatment of all neurological disease has been revolutionized by better diagnostic technologies, therapeutic drugs and personalized surgical techniques, which means that outcomes for people like Betty are improving. I am sure that in the future the manipulation of the microbiome will be part of this therapeutic arsenal.

This is the century of the gut–brain axis; we are at last becoming conscious of our microbiomes. Freud would have loved it.

8. How to Live to One Hundred

Legend has it that in 1513 the Spanish explorer and conquistador Juan Ponce de León set off from Puerto Rico in search of the island of Bimini – and accidentally found Florida in the process. Bimini, Ponce de Leon had been told, was the site of the fountain of youth: a pool of water enveloped by gigantic magnolias that had for centuries been kept 'as fresh as violets' by the magical properties of the fountain's water. The water would have contained protozoa, phage, fungi and bacteria and, for my money, that symbiosis is more likely to have given the magnolias their long life than any kind of magic. We don't know whether Juan ever did find the island, though if he did, Bimini's youth-giving properties can't have been all they were cracked up to be: Ponce de León was dead by the age of forty-seven. Bacteria were responsible; he died from sepsis after being wounded by an arrow to the thigh.

The fountain of youth is a recurring feature in the human imagination, first referred to by Herodotus in the fifth century BC and reinvented by successive generations. It's a concept beloved by billionaires who inevitably all come to the same realization: all the money in the world cannot change the one inevitability common to all life forms. Most recently, Jeff Bezos, founder of Amazon, invested in a start-up called Altos, which aims to reverse-engineer the ageing process. The powerful will always keep looking for Bimini.

Scientists have also long wrestled with questions of ageing and mortality – perhaps none more so than the Russian immunologist Élie Metchnikoff, for whom the gut was a 'vestigial cesspool' of fermenting microbes at the centre of this ancient struggle. His verdict is nothing short of damning:

> The intestinal flora of man appears to be the principal cause of the short duration of human life, and the human conscience has been aroused to rectify this injustice.[1]

Metchnikoff urged people to avoid eating raw food wherever possible, convinced it was a source of wild microbes and toxins. He even advocated surgery to remove the entire colon so that the gut microbiome could be completely eradicated.

If you want to live a long and healthy life you don't need your colon removed; in fact I highly recommend that you hang on to it. When it comes to extending your life, you don't even need to know the first thing about the microbiome. Just follow these four simple rules:

1. Don't smoke (or vape).
2. Don't be overweight.
3. Don't ride a motorbike.
4. Live in a country with an affordable healthcare system.

The problem is that not all of these things are choices. Most overweight people don't choose to be obese; we cannot choose where we are born; smoking is chemically addictive; and economics or poverty dictates where many people live. If you are a lean, non-smoking pedestrian reading this in Japan, then congratulations: there is an excellent statistical chance you will make it to at least 84.2 years of age. If you are reading this while smoking a cigarette, waiting for your hamburger to turn up in a motorcycle café in Charleston, West Virginia, then it is more likely that you will be alive for 74.4 years or less. West Virginia has the lowest life expectancy in the US. But don't feel bad, as there are a million ways to die and, far more importantly, there are a million ways to live. Riding a motorbike is exhilarating and there is no better place to do this than on the country roads of West Virginia, which will always take you home.

I'm a strong advocate of ageing disgracefully and of the importance of fun. But I will also concede that, as a species, we are very bad at adopting simple behaviours that are definitively proven to lengthen our healthy lifespans. Part of the reason for this is that we don't really need to. Modern medicine compensates for our poor decision-making and, if you can afford it, you will now live longer than you were designed to. The phenomenon of the ageing population is both pronounced and historically unprecedented. In the next four decades, approximately 22 per cent of the global population will be over the age of sixty: a jump from 800 million to two billion people.[2] And for the first time in human history, people aged sixty-five and older outnumber children aged five and under.

Although population data suggest that elderly people are healthier than ever, wide gaps in life expectancy between the worst- (Sierra Leone) and best-performing (Japan) countries remain, with a range of thirty-six years for life expectancy at birth. In the developed world, the challenge is not how to live a long life, but how to enjoy a good *quality* of life, whereas in the developing world survival remains the priority.

Ageing is a complex biological process that we all experience differently, but it's defined as progressive functional deterioration associated with frailty, disease and death.[3] Older people possess multiple coexistent diseases known as comorbidities, and these manifest through a loss of physiological function and the broad geriatric syndromes of impaired cognition, continence, gait and balance.[4] The gut microbiome plays a critical role in defining these anatomical and physiological responses to ageing. It influences digestive functions, bone density, neuronal activity, immunity and our resistance to pathogen infections. It also plays a central role in the effectiveness of preventative strategies or drug treatments that are widely used to treat the diseases of old age. Anyone trying to shape policy on healthy ageing does need to know about the gut microbiome after all.

The GEM line

While there are an infinite number of factors that will determine how we age, we can helpfully group them into three main forces: the genome, the exposome and the microbiome. These are the three forces that shape not only our risk of disease, but also our experience of wellness and healthy ageing. Think of them as the axes of a three-dimensional graph on which we can plot the progress of our lives (Figure 1).

First up is the genome: all of the genetic information that makes up our bodies, and a major player in our ability to grow, function and age normally. For a minority of us, genetic mutations that we inherit will set our risk of getting a disease or experiencing an early death. But only 25 per cent of our health and function in older age is determined by our genes,[5] with the remaining 75 per cent strongly affected by the cumulative effect of our behaviours and environment. That is because environmental toxins and harmful agents inevitably lead to damage to our genetic code. 'Epigenetics' describes how cells control gene activity without permanently mutating or changing the DNA sequence itself. This is essentially a genetic expression of gene–environment interactions, which alters how genes are turned on and off without permanent damage to the genetic code itself. As you would expect, epigenetic gene modifications are also a major determinant of our risk of disease.

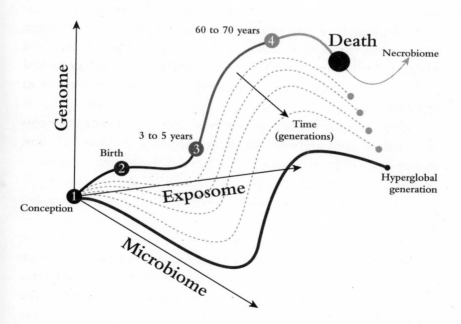

Figure 1. The GEM line – a theoretical description of how our health is defined by our genome, exposome and microbiome and how it has evolved over successive generations to influence our risk of chronic disease.

Next, the exposome: all the environmental forces that interact with our genome over the course of our lives. This includes the place you live, the sea you swim in and the soil you stand on, the food you eat, the people you kiss, the cigarettes you (do or don't) smoke, the car you drive, the social-media platforms you use, the family you interact with, the church you go to, the job you have, the medicines you take and the social class you were born into. Unlike the stable genome, the exposome is dynamic – it's influenced by the many chemicals, proteins, toxins, plastics, sunlight and gases that you come into contact with. Your gut is your greatest environmental sensing organ, and it will taste, smell and ingest all of these things. In the course of a single day the intestinal immune system is exposed to an antigen load greater than that encountered by the systemic immune system during an entire lifetime. Which is why it has a complex anatomical and physiological architecture that allows it to function and to educate the innate immune response – a major determinant of both your health and how you age. The gastrointestinal tract represents the largest immunological organ in terms of numbers of lymphocytes (white blood cells that fight infection) and at any given time the gut-associated lymphoid tissue (GALT) will harbour up to 20 per cent of the total lymphocytes in the body.

Finally, the microbiome: the collection of microscopic life forms across all sites in and on your body, including all forms of symbionts, amensalists and pathogens. The microbiome has an unstable and dynamic construct that has its greatest influence on the development of our lifeline before the age of three, and during this period it remains under the influence of the parental microbiome. After this point it develops into a more stable ecology representative of the adult gut, remaining relatively constant throughout adult life until the age of around sixty or seventy, when it begins to decline.

The three life forces of genes, exposome and microbiome – or GEM – continuously act on us and interact with each other. Rather than considering our lifeline, we can start thinking about our GEM line.

Our GEM line is what scientists call our 'phenome', which

describes all the measurable physical and chemical characteristics of the human body. The influence of these characteristics varies as we age, so the GEM line isn't straight, but curved and distorted as it wiggles along. It maintains internal stability while adjusting to changing external conditions and is able to adapt through growth and reorganization. A hallmark of microbiome health is its resilience in dealing with fluctuations in the exposome or genome. For this, a healthy microbiome needs to have a large functional reserve capacity, which comes through diversity. Healthy ageing is therefore not simply the absence of disease. It is better thought of as a series of dynamic features that maintain a finely balanced equilibrium along this GEM line – at all scales of cellular life and across all compartments of the body.

This is complicated further by the fact that the ageing body is continuously turning itself over, replacing damaged cells and regenerating new ones, reconstructing injured defences that keep out molecules and microbes that wish us harm. Tissue must sometimes be destroyed for our own good, and then perfectly reconstructed by our genome. The rate of regeneration varies dramatically across organs like the skin, the lung, the gut (fast) and nerves (very slow). Programmed cell death is called apoptosis – it's a process under the control of suicide genes, but it can be usurped by the microbiome. The replacement of dead cells by stem cells (ones that have the capacity to grow into any type of cell) is also not easy, as it must maintain its size, its genomic integrity and its epigenetic identity – three critical features that we tend to lose with old age.[6] Unsurprisingly, this process is also vulnerable to microbial interference. The GEM line doesn't just undulate with the pattern of your sleep, your menstrual cycle (if you have one) or environmentally induced hormonal shifts like a midlife crisis. It vibrates in tune with every oscillation along each of the three axes of life. It exists in a quantum state that can be simultaneously healthy and harmful, because the consequence of these fluctuations is inflammation, which is the master regulator of biological ageing.

There comes a point – or several points – in our lives when one of the axes will exert such a significant force on us that the GEM-line trajectory shifts more dramatically, and this will manifest itself as a 'disease'. And for all of us, at a certain moment, our lifeline will stop for ever in the still blackness, while microbial life will keep going, decomposing our bodies back into the earth.

Can we predict which trajectory any one of us is on and, more importantly, is there anything we can do about it? You cannot easily change your genes (yet), but you can change your exposome and you can engineer the microbiome. In our quest for eternal life, the microbiome represents an important target when we are too lazy, stubborn, poor or manipulated to change our behaviour.

Currently we can only make broad predictions about where and when an individual's GEM line is likely to end, because we have no historical measures of these complicated molecular interactions upon which we can build more accurate models. Genetically, however, we're very similar; our behaviours are broadly predictable; and particular cohorts of humans are likely to experience similar exposomes. With this as a starting point, and excluding any accidents, epidemiologists can use crude measures to predict how long we are likely to live. Really good epidemiological models can even provide us with helpful advice on how to stay healthy. But your GEM line is as unique to you as your fingerprint, and these recommendations can't account for your individual systems biology or the impact of your microbiome on your health. We lack precision or personalization in our mathematical models, and until we include the microbiome in our calculations these statistical tools will remain blunt.

The GEM line allows us to conceive of how these three dominant forces influence our health over several generations. Young people, born during the age of hyperglobalization that started in the 1990s with the rapid growth in economic, cultural and political globalization, have experienced such a profound disruption to their microbiome and exposome that their risk of disorders of the immune system has been dramatically altered. Their trajectory is very different from those of us born before this era (Figure 1, page 127), and this in turn bears little resemblance to someone who was born before the Second World War.

Of course, not all exposome drivers of disease are equal, and some work across microbiomes in different parts of our body. Smoking gives us a particularly good example of this.

The gut–lung axis

Just like the gut, the lung is first colonized during birth, and then by very small aspirations as we breathe in or cough. This passive process takes place largely during sleep, when our oral and pharyngeal muscles relax – which is in part why those of us who snore have different lung microbiomes from those who don't. Studies sampling the lungs of sufferers of obstructive sleep apnoea have shown that snorers have more Proteobacteria and Fusobacteria and fewer Firmicutes.[7]

In healthy individuals the lung microbiome is noticeably different between the upper airways – made up of the big, whistling cartilaginous tubes called bronchi – and the lower respiratory tract, made up of soft and squidgy alveoli. The lower respiratory tract contains *Streptococcus*, *Prevotella* and *Veillonella*.[8] In chronic infections of the lung, these bacteria are often found in large clusters, called biofilms, which are much harder to physically cough up or remove and more resistant to antibiotics. Your nasal microbial community also matters: in healthy individuals this is closer to the skin microbiota (Firmicutes and Actinobacteria phyla are the most prevalent), and the nasal microbiome is incredibly important in stopping pathogens from infecting us.

Smoking is the biggest modifier of the lung microbiome. And please don't think that vapes are any better for the lung, for the gut or for you. Filling our lungs with propylene glycol carries many of the same risks as exposure to cigarette smoke, and it makes already-harmful bacteria more dangerous.[9] Some 1.3 billion people worldwide use tobacco products, mostly in the form of cigarettes, with tobacco smoking-related diseases resulting in at least eight million global deaths per year.[10] Smoking is the principal cause of lung cancer and chronic obstructive pulmonary disease (COPD). But smoking is also strongly linked to the causes of a variety of nasty systemic diseases, including cardiovascular disease, inflammatory bowel disease and various forms of cancer, like bladder cancer. Tobacco smoke itself is highly toxic, with high carcinogenic potential. But it goes deeper than that: tobacco smoking affects the peripheral immune system, which is where the microbiome comes in.

Smoking changes the lung microbiome directly, although we're still figuring out the exact mechanisms. Each brand of cigarettes has its own microbiome, and European Union cigarettes contain up to fifteen different classes of bacteria, ranging from soil microorganisms and commensals to potential human pathogens.[11] But it's more likely that smoking changes the microbiome by altering the 'microenvironment' of the lung. It alters the activity of the cilia that transport mucus along its lining and influences mucosal oxygen, pH and acid production; and all of these, in turn, change our lung ecology. Cigarette smokers have a different lung community structure, with greater abundances of *Prevotella*, *Veillonella*, *Streptococcus* and *Actinomyces* and of opportunistic pathogens.[12] The more they smoke and the longer they smoke, the greater the changes to the lung microbiome, and the harder it is to go back to a normal ecology. Even the commensal viruses in the lower respiratory tract are different in smokers and non-smokers.

Smokers don't just have different lung microbiomes. All of their body's ecology is different from non-smokers. For example, smokers have a very perturbed gut ecosystem and lower relative abundance of *Prevotella* and *Neisseria* spp. and an increased relative abundance of Firmicutes, principally *Streptococcus* spp. and *Veillonella* spp., along with the genus *Rothia* (Actinobacteria) in the upper gastrointestinal tract. Smokers also have more Bacteroidetes and decreased abundances of Firmicutes and Proteobacteria in their gut compared to those who have never smoked.[13] These differences matter because microbiome–immune interactions are driven by the gut. They might explain, for example, why smoking is such a strong risk factor for the inflammatory bowel disease called Crohn's, and why it influences an individual's risk of bowel cancer. A lot of the harm caused by smoking comes from sustained changes to both the gut and the lung microbiomes, which in turn affect the microbiome–immune system.

Jeanne Calment may have been the oldest woman who ever lived.[14] At the time of her death, she was apparently aged 122 years and 164 days, and only stopped smoking aged 117. However old she really was, she had lived an extraordinary life, and my guess is that she had a very healthy microbiome. But my recommendation, to any smokers reading this, is still to stop smoking!

The ageing microbiome

The social and physiological changes of age have a profound effect on the intestine. We lose our teeth and have trouble chewing; our stomachs thin and we're more at risk of infection; our gut hormones become less active and we start developing chronic diseases of the intestine ... And that's before we consider social isolation, reduced mobility and malnutrition.[15] These factors all change the microbiome.

We haven't yet defined exactly how old we are when the gut microbiome starts to age – but it's generally accepted to be somewhere between the ages of seventy and eighty. As we get older, the relative proportions of the Firmicutes and the Bacteroidetes change, and generally (but not always) older people have a greater abundance of Bacteroidetes, while young adults have higher abundances of Firmicutes.[16] It is now, however, the proportions of the bacteria that change, and our microbiome becomes more individualized because it has had to deal with a lifetime of our habitual behaviours and peccadilloes. The elderly microbiome therefore has more distinct taxonomic and functional compositions than that of other adult age groups. These observations vary, though, depending on where elderly participants are living when they have their microbiomes sequenced. This is important, because it means that the elderly gut is harder to engineer to improve health and there are other important implications for this observation, and you should bear this in mind before you ship your elderly relative off to a care home.

A study by researchers from Cork, Ireland, studied the faecal microbiota composition of 178 elderly subjects and found that these correlated closely with where the person was being cared for.[17] People's gut microbiomes changed depending on whether they were in a day hospital, in rehabilitation, in long-term residential care or in the community. These differences were influenced by diet and by the sharing of microbes between residents, once the participant had been admitted into a particular environment. The significance of this is that the elderly microbiome has less resilience than that of young people and lacks plasticity.

The changing microbiota seen within these groups significantly correlated with measures of comorbidity, nutritional status and markers of inflammation. The microbiota of people in long-stay care was significantly less diverse than that of community dwellers, and this correlated with increased frailty. Anyone who works with older people understands the importance of nutrition, but frailty in our elderly population in residential homes might, in part, be influenced by the gut microbiome – and personalized nutritional strategies that account for both the individual microbiome and the impact of the changing environment would be an important way to minimize the risk of frailty and improve the function of the brain. The same applies to any institutionalized populations, such as schools or prisons, where diets are aligned and microbes are freely shared between classmates and inmates.

We don't know yet whether it's ageing that changes the microbiome, or changes to the microbiome that age us. Pockets of centenarians can be found living in discrete geographical regions across the globe – including Okinawa in Japan, Sardinia, Nicoya in Costa Rica and Icaria in Greece – which have been named 'blue zones' by the author Dan Buettner, because people who reside within these regions appear to live longer. From a microbiome perspective, this is a false epidemiological construct, as longevity has not been limited to these specific regions over time; for a start, Metchnikoff would have added a fifth zone, Bulgaria – although sadly this would not be true today. He observed that Bulgarians drank large amounts of a fermented yoghurt drink. From this, he isolated the probiotic *Lactobacillus bulgaricus* and argued that it was responsible for Bulgarians living such long lives, even though evidence suggested it couldn't be cultured in the gut. In truth, blue zones are subject to confounding exceptions. London could be argued to contain its own blue zones, as we see massive variation between its boroughs in years lived, based on factors such as social status, mobility, social inclusion, diet, stress, purpose and exposure (or otherwise) to pollutants and toxins like cigarettes and alcohol – all associated with the exact postcode that someone happens to live in.

It's possible that there are members of the gut microbiota specific to people who live past 100 that promote health, and help resist pathogens and other environmental stressors that speed up the rate at which we age. A recent analysis of Japanese centenarians identified some bacteria taxa that could only be found in the intestines of people over the age of 100. One of the most-enriched species in centenarians was *Clostridium scindens*, known to possess a relatively rare enzyme called 7α-dehydroxylase that's needed to metabolize bile. Bile – the thick, green digestive fluid made in the liver and stored in the gall bladder – is a potent bodily fluid with a multitude of functions,[18] and bile acids are linked to many other changes in the ageing process. Bile influences muscle-thinning, obesity, cardiovascular and cancer disease risk, and even the risk of neurodegenerative disease. Centenarians, it seems, have a distinct gut microbiome that is really good at generating unique bile acids, which limit inflammation in the gut.

Gut bacteria are also able to produce molecules in the gut that can directly impact ageing. An intriguing experiment at New York University illustrated how one small molecule produced by one bacterium can dramatically affect the physiology and longevity of a completely different organism.[19] The experiment used the nematode *Caenorhabditis elegans* as an experimental ageing model. These worms are unable to produce nitric oxide – a gas that, in humans, helps increase blood flow to the gut, transmits nerve signals and regulates immune function. As we age, nitric-oxide levels in the gut decline. While the worms can't produce nitric oxide on their own, scientists wanted to find out what would happen if they were grown with the microbes (*Bacillus subtilis*) that could do so on their behalf. When they did so, the worms' lifespan increased by nearly 15 per cent, to about two weeks. The nitric-oxide gas switched on two master genes in the worm (*hsf-1* and *daf-16*) that resulted in a greater resistance to stress and, hence, a longer life. The question now is whether the same bacteria have this impact in the human gut. We're yet to find out.

Inflamm-ageing

The molecular cross-talk that constantly goes on between gut microbiota and the immune system enables humans to tolerate and influence gut ecology, but it also leads to a low-grade inflammatory state. One of the most important effects of ageing is that the adaptive and innate immune systems in the gut lose their power to influence this cross-talk.[20] As we get older, the body finds it harder to balance pro-inflammatory and anti-inflammatory pathways, which profoundly influences the ageing process. In practice, this process occurs at different rates in different organs, and it leads to what's known as the 'mosaic of ageing'.

'Inflamm-ageing' describes the consequences of a person's ability to cope with the increasing burden of environmental drivers of inflammation.[21] And it explains why our GEM line shifts as we get older. The longer we live, the longer the human immune system has to face antigen exposure – several decades more than in our recent evolutionary past. Centenarians manage a complex balance of pro-inflammatory and anti-inflammatory pathways, whose net result is a slower, more limited and balanced development of inflamm-ageing. Disruptions to this balance go some way towards explaining the rise in non-communicable diseases associated with ageing and inflammation – be they atherosclerosis, cardiovascular diseases, type II diabetes, metabolic syndrome, osteoporosis or cognitive decline and frailty.

As we get older our immune system is less able to respond appropriately when it's called to action. We become 'immune-senescent' as the immune system gradually degrades – a large number of environmental stressors may trigger the senescence state, but it seems that it is susceptible to dietary manipulation and microbiome-host interactions.[22] The result is that we are less able to mount an immune defence when we really need one – for example, when we need to fight a pathogen (like Ray, see page xvi) or repair an injury. And what is the most critical organ where inflammation influences the quality of our ageing life? The answer is obvious: it is the brain.

How we are losing our minds

As we get older, the blood–brain barrier begins to tire. With this, microbes, their toxins and proteins begin to cross over into our subconscious and they mist our memories.

Almost 80 per cent of dementia cases are caused by Alzheimer's disease. This occurs when proteins called amyloid plaques and neurofibrillary tangles coalesce in the nerves that form the memory centre of the brain.[23] Incredible advances have been made in our understanding of how these proteins form but, for now, treatment options are limited to slowing down the progress of the disease and relieving its symptoms.

Other neurodegenerative disorders, such as Parkinson's disease (PD), Lewy Body disease (LBD) and Multiple System Atrophy (MSA) are also caused by the production of an amyloid protein called α-Synuclein. Many individuals diagnosed with these conditions will complain of constipation or altered bowel habit before the decline in their central nervous system becomes apparent. This may simply reflect the neurological changes that influence the second brain in the gut, but it may also point to something else.

People with Alzheimer's have a different gut microbiome from people of the same age and gender who do not have dementia. For example, they have been found to have increased abundances of Proteobacteria and Bacteroidetes, and relatively lower abundances of the phylum Firmicutes and the genus *Bifidobacterium*.[24] Scientists are now trying to establish the relevance of this observation, and it is clear that the microbiota have an important role to play. For instance, germ-free mice engineered to develop Alzheimer's don't develop it unless their gut is colonized with bacteria given by FMT from healthy animals with the same genetic predisposition for this form of dementia.[25] And when mice undergo FMT from humans with Alzheimer's, this is also enough to produce the same cognitive decline through the formation of amyloid in the brain.

In fact bacteria that live within the gut are able to produce proteins that have very similar structures to amyloid proteins, and there is even evidence in rats that specific strains of *E. coli* are able to influence the deposition of α-Synuclein in the brain.[26] It is not known for sure how these bacteria actually transport these proteins to the brain, but the most likely route is via the vagus nerve. This is because mice injected with α-Synuclein (the protein that causes dementia) do not get the condition if the vagus nerve is cut, and this observation holds firm in humans too.[27]

The microbiome also appears to be important in causing equally devastating neurodegenerative diseases that affect young people. Amyotrophic lateral sclerosis (ALS) is a motor-neurone disease, meaning it destroys the cells that control skeletal muscle activity such as walking, breathing, speaking and swallowing. A recent study from the Weizmann Institute of Science reported that mice that were genetically engineered to get ALS experienced more aggressive disease if they were given antibiotics or grown in a germ-free environment.[28] Just eleven bacterial species seemed to explain this observation, so the researchers then selectively transplanted each of these bacteria into the genetically engineered ALS germ-free mice. *Ruminococcus torques* and *Parabacteroides distasonis* exacerbated the disease course, while the bacterium *Akkermansia muciniphila* was protective. Part of the reason it was beneficial was that it promoted the concentration of an essential vitamin called nicotinamide (vitamin B_3, found in fish, poultry, nuts, legumes, eggs and cereal grains) in the central nervous system. Supplementing the mice's diet with nicotinamide improved their motor functions. In humans, researchers found that ALS sufferers had similar microbiome and metabolite configurations that caused reduced levels of nicotinamide in their cerebrospinal fluids, when compared to family or household members.

This exciting work suggests that we can target the functions of our gut microbes over our entire lifetime to improve the health of our brain – and that it might be possible to prevent our cognitive and motor functions from declining, by optimizing the gut microbiome. This research gives us real hope that in the near future it will be possible to engineer the microbiome to develop the treatments we're currently lacking for neurodegenerative diseases.

The meta hypothesis

In the first book of his magnum opus the *Metamorphoses*, the Roman poet Ovid described the four ages of man, and in doing so gave us a useful framework for understanding the importance of the microbiome to human health (this can also be seen in Figure 1 on page 127).

The first age refers to the *maternal microbiome*. Here, changes in the microbial orchestral signalling between a mother and her developing baby cause epigenetic modifications in the developing organs and immune system. This sets the course for the whole of our lives, our brain development, behaviour, personality and our risk of chronic diseases.

The second age takes place *in the first three to five years of life*; within this period, the most critical time is the first 100 days. If the seeding and dispersal of the developing microbiome are interrupted, then the immune and metabolic systems can't complete the programming of the systemic organ axes (brain, liver, lung, bone and heart). Our risk of disease changes once more, and our risk of allergy, atopy, autoimmune disease and mental-health problems becomes more defined.

The third age is the broadest and takes place *between the ages of five and seventy*. During this phase the ecological structure of the microbiome is relatively stable, but its functions can oscillate on an hourly and daily basis. Its chemical outputs are influenced by our diet, drug and xenometabolite (non-host metabolite) interactions, which in turn determine our risk of diseases such as obesity and cancer. It's hardest to modify the microbiome for health benefits during this stage, which is at its most fixed.

The fourth age takes place *between the age of seventy* (assuming we get that far) *and our end of life*. The microbiome is once more dynamic in its structure, influencing the balance between inflamm-ageing and immune-senescence. This in turn determines our cognitive decline, mobility and frailty, the quality of the end of our life and our death. This is as important to our experience of life as our birth is. The microbiome even maintains its influence after death, through the decomposition of our flesh (this is known as the necrobiome).

The metamorphoses model illustrates that our microbiome is not constant, and its requirements change as we enter each age of life. We all arrive by different routes into the world – in different cities and villages, with varying levels of pollution and antibiotic resistance, family sizes and pets. Some of these things we can control, and others we cannot. Each of these microbiome–exposome–genome interactions has to be understood along timelines that cross generations. It explains why East Germans and West Germans had different risks of allergy and atopy over time, despite the fact that the populations were genetically similar and geographically co-located. It explains why millennials have a risk of bowel cancer that is four times that of baby-boomers. It explains the explosion of neurodevelopmental conditions, obesity, cardiometabolic disease and autoimmune disease. It explains why even vegans get sick sometimes. This 'meta hypothesis', as it's known, doesn't necessarily imply that we can't alter our risk of chronic disease at any one of these stages. But it does mean that the later in our lives we understand this principle, the harder we'll have to work to correct this risk.

Life after death

We do not die well in the twenty-first century. We don't get to choose where or how we die, and I've watched too many people die in hospital away from their families. Death has not only become rarer in developed societies, it has also become sanitized, and we've become physically and socially removed from it. Yet life is so precious, and we don't go gently into the good night; we 'fight' cancer and we win battles against sepsis, and we do whatever it takes to remain present, even when we aren't. And if we want to give up – if we don't want to fight or we express a wish not to suffer – we're denied that option. Society doesn't permit any other choice but to endure. It is not possible to opt out of life.

In 2020, COVID-19 changed this. Death was everywhere, and we had to deal with it on an unimaginable scale. Living became the choice that we were unable to make. The horror was (and has been, at the time of writing) overwhelming; palliative care services were overrun, and the sick and frail were often left to die alone at home with little or no support. The acute trauma of bereavement was exacerbated for devastated families who were forcibly separated from their loved ones. Loneliness could be the most crippling of all of the symptoms of COVID-19.

I would not choose to live for ever. If billionaires can find a way to push back the tide for just a little bit and come up with a way of dodging death, the method will undoubtedly be prohibitively expensive and the healthcare divide will only widen. How much better it would be if they chose to invest their extraordinary resources in finding new ways to lift us all up to the same living standard, or even in improving the quality of our deaths.

There is no secret to living a long and healthy life; place an emphasis on mobility and nutrition, and give the gut–brain axis a sense of purpose. Avoiding the care home in the first place may be the best strategy when it comes to ageing well. Intermittently restricting our food intake might also be one of the single best ways of extending our lives. Depriving our microbes of nutrients triggers a series of events that changes the body's resistance to insulin and its metabolic rate. It also initiates a cellular process called autophagy that regenerates nutrients from macromolecules and clears damaged material from the cellular environment. Unsurprisingly, the biggest direct impact of intermittent fasting is on the gut microbiome. It works as an anti-inflammatory diet that enriches bacteria in the gut that are generally important for health, such as *Lactobacillus* and *Bifidobacterium*, while reducing the abundance of poor health phylotypes, such as *Helicobacter*.

The microbiome is as important for our last hundred days as it is for our first, and I feel optimistic about the part microbes will play in my death. As our bodies enter the final stages of life, our microbiomes influence how our metabolism slows, how our barriers fail and how the mind mists. It influences the mucus we produce that sits in our airways, and our sensation of breathlessness. It will influence the pain we feel and the effectiveness of the drugs that will keep us 'comfortable'. If we understand the microbiome, we can improve the quality of palliative care and the dignity with which we die. I hope by the time I die (whenever my GEM line decides that is) this is more deeply understood.

For me, death will be the end, but for my microbes it will be simply another opportunity. The saprotrophs and necrobiome will start their work and another world will come to life within me, sharing my molecular memories and my atoms with the earth and the soil that needs them. It is what Elton John would call the Circle of Life. It leaves me feeling wonderfully optimistic and thankful, knowing that some part of me will be recycled into the microbial cosmos and beyond.

PART TWO

The Exposome

9. The Global Microbiome

When my grandfather died in the 1960s, my grandmother left the United Kingdom. She became a nomad, exploring the world, before settling in Tortola, one of the British Virgin Islands. There she busied herself in island politics while doing her best to cope with loss and a rapidly changing world. As very young children in the early 1980s, my brother and I were taken on epic transcontinental journeys to visit her. These required jets, rickety propeller-driven Dakotas and loud wooden motorboats, all of which could have fallen from the sky or sunk into the shark-infested waters. She would greet us at the airport in her rusty grey VW Beetle and career excitedly along the dark hilltop roads lined with banana trees, chatting all the while. She loosely steered with one hand and with the other would hold a cigarette, in a distinctive black-and-gold cigarette holder.

It's difficult to convey just how fabulous this was for a six-year-old. I would wake disorientated and overexcited in the bright dawn of her small house, which sat among the mangroves on its own tiny beach. One of my favourite pastimes was to lie on her wooden jetty, staring into the still, glassy water, and examine the alien life forms that did not exist in Brent: huge fish, bizarrely shaped and brightly coloured fish, starfish, crabs and even eels. They would stare back at me, equally bemused. Every day my grandmother would swim among them, oblivious to the sharks. The small island is still there, but the mangroves have gone, and so have the fish.

The fourth age of man

We often think about the consequences of climate change in terms of the loss of biodiversity of trees and plants or fish, insects and animals who have faces, personalities and Instagram accounts. But the greatest biomass and biodiversity on our planet comes from microbes. We can't even see most of them, because 70 per cent of all bacteria and archaea live at least three miles underground, with a total biomass estimated to be about twenty-three billion tonnes.[1] Bear in mind there are only about 385 million tonnes of humans.

Microbes are essential for the survival of the planet, and the delicate microbial communities found in the soil, the atmosphere and the sea are being fundamentally altered by climate change. Their importance is so great that in 2019 the Alliance of World Scientists put humanity on notice, warning us that microbes are responsible for recycling the fundamental elements essential for life on Earth, namely carbon, nitrogen, hydrogen, oxygen, phosphorus and sulphur.[2] They've developed an array of methods for using these elements as their energy sources, cycling them continuously through you and me and our environment. When we destroy our environment, we also damage our own internal ecosystem.

Industrial spills associated with extracting, transporting and consuming oil total on average 1.3 million tonnes per year. When they happen, microscopic life forms suffer like the rest of us. The *Deepwater Horizon* oil spill was one of the worst environmental catastrophes in history, dumping a massive 4.9 million barrels of oil and 250,000 tonnes of natural gas into the Gulf of Mexico over eighty-six days.[3] The impact on marine ecosystems has been catastrophic; its destruction of the oceanic microbiome is less well understood but of equal importance. More than 2,000 historical shipwrecks spanning 500 years of history rest on the Gulf of Mexico seabed. The German U-boat *U-166* and the wooden-hulled sailing vessel known as the *Mardi Gras* wreck, both in the Mississippi Canyon leasing area, were exposed to deposited oil. Analysis of the shipwreck microbiomes showed microbial ecological changes, with significant increases in Piscirickettsiaceae sequences in surface sediments, and reduced biodiversity.[4] If the microbiology of the ocean floor changes like this, it influences the entire marine food chain, which depends on accessing a diverse and stable microscopic banquet. Ultimately, this influences the abundance of seafood that we rely on, but also its quality and safety.

Bacteria might, however, offer a solution for managing future environmental catastrophes. During the *Deepwater Horizon* spill, colonies of the bacteria *Marinobacter* and *Alcanivorax* found in microalgae bloomed as they began to feed on the petrochemical hydrocarbons.[5] This change was so dramatic that these hydrocarbon-munching microbes suddenly made up almost 90 per cent of the microbial ecosystem in the local contaminated waters. We could now potentially engineer these microbes into oceanic probiotics able to consume toxic spills in a safer manner than dispersants, which are chemicals that break oil into small molecules. Of course massive blooms of genetically engineered probiotics bring their own challenges, and the best strategy might be not to rely on oil in the first place.

Petrochemicals are used to make thousands of products, perhaps the most important of which is plastic. We have carefully bubble-wrapped ourselves in so much of this synthetic material that we now exist in an entirely new epoch, known as the 'plasticene'.[6] This has important implications for our health, not least because we inadvertently consume about 5g of it per week. An apple a day keeps the doctor away, but it also contains an average of 195,500 plastic particles per gram, while pears average around 189,500 plastic particles per gram.[7] Plastics also enter our food chain via the sea, where they're consumed and filtered by oceanic plankton, bottom-feeders and crustacea.

Because plastic has become part of our environment, it's changing our microbiology – and planetary microbial systems are having to adapt. Nanoplastics cause harm either through direct mechanical disruption to the gut and its barrier or by altering the workings of our endocrine systems. Mice that have microplastic polystyrene particles placed in their drinking water lose diversity in the gut microbiome within twenty-eight days.[8] Some members of the colonic microbiota also stick to the surface of microplastics, eventually forming biofilms. We're also seeing evidence that microplastics influence the functioning of our immune system – either directly or through changes to the microbiome[9] – with important implications for our risk of chronic diseases. In male mice, for example, microplastic consumption causes infertility through a gut microbiome-mediated change in the immune system.[10]

Only 9 per cent of the plastic ever produced has been recycled, and every plastic bottle takes 100 years to biodegrade: we need a solution to plastic pollution. The main chemical component of plastic is polyethylene terephthalate (PET), used for its extreme durability, and bacteria may well be our best hope of dealing with its environmental consequences. Along with yeasts and fungi, *Ideonella sakaiensis* was isolated from sediment at a bottle-recycling facility in Osaka in 2016.[11] This bacterium has evolved to break down and metabolize plastic by developing enzymes called hydrolases. The particular enzyme responsible for this process has now been isolated and mutated; the ambition is to engineer ecological microbiomes to dissolve the plastic waste mountain. Whether this can be turned into an efficient and cost-effective strategy remains to be seen.

Feeding into and exacerbating the climate crisis is our addiction to consumption. In 2020, human-made mass reached about 1.1 teratonnes, exceeding the total global biomass on the planet.[12] This makes humans the greatest hoarders ever to have lived; if we continue in this vein, anthropogenic mass is projected to triple global biomass by 2040. To reach this milestone of manufacturing we've destroyed large swathes of the planet's natural landscapes, transforming over 70 per cent of the planet's land surfaces, with a dramatic impact on our planetary microbiome. Since the first agricultural revolution in 10,000 BC, humanity has roughly halved the biomass of plants, and has disproportionately grown domesticated species by converting 40 per cent of our land mass to unsustainable farming practices. We use about three-quarters of freshwater supplies, and we are busily setting fire to the planet's most precious and biodiverse ecosystems. Industrialized farming has fundamentally altered the 'agribiome' through soil degradation, contamination, deforestation and massive biodiversity loss. Just as it's transforming our understanding of humans, so metagenomic analysis of farming soil populations is giving us new insights into previously unknown species of bacteria and fungi, which can, we hope, be leveraged to promote crop growth and resilience and to reduce our reliance on pesticides.

Our gut microbes have a vast capacity to metabolize environmental chemicals such as pesticides through a core of enzymatic families. The bacteria in our gut determine the toxicity of the pollutants we consume, which is why they affect us all differently. Environmental contaminants also change the composition and functions of these bacteria in the gut and cause chronic diseases. Residents of Greater Baltimore, for example, have been exposed to unusually high levels of Chlorpyrifos, one of the most common organophosphorus pesticides. It caused so many health-related problems that the city banned pesticides containing this agent in 2020. The damage inflicted on the gut and its microbiome by this toxin has been linked to the high rates of obesity and diabetes among the city's inhabitants.[13] Gut microbiota are a hugely underestimated intermediary in the toxicity of environmental contaminants.

Bacteria are the masters of adaption and, by sharing their gene mutations, they're capable of evolving to new and toxic climates much faster than we are. This battle for survival is taking place outside and *inside* our bodies, which means that climate change is happening in you, right now. If your gut were a reef in the British Virgin Islands, it would be bleached and dull; and if it were an Amazon rainforest, it would be on fire.

The global microbiome

The human genome is incredibly similar between people, regardless of our skin colour, caste or social status, while the microbiome varies greatly. Our genes only have a minor role in determining the make-up of our microbiome[14] – a recent study involving 2,252 twins suggests that this could be as low as 1.9 per cent.[15] The differences that we see in the microbiome across various ethnic groups are not, therefore, down to our genetic differences, but occur as a result of environmental drivers, including diet, culture and climate, and some of these drivers are subject to abuse. Race and caste are social constructs, and they are not a basis for microbiome science. But ethnic microbiome variations are important to understand for two reasons: first, they tell us why some social groups are especially vulnerable to disease; second, they tell us how we can modify the exposome or the environment to do something about it.

Although broad abundances of the four largest families or phyla of the gut microbiome are present in people from across the world, their relative abundances vary greatly. For example, at the phyla level the rural African gut is dominated by *Prevotella*, while in the Western gut *Bacteroides* predominates.[16] These global variations in gut ecology map closely on to geographical location – and this has revealed entire bacterial groups abundant in industrialized individuals that are almost absent in non-industrialized individuals. In fact, Western microbiomes consist of 15–30 per cent fewer species of bacteria than non-Western microbiomes. At the species level, this global variation becomes even more pronounced, and it is remarkable. For example, separate subspecies of the same bacteria, such as *Eubacterium rectale* and *Prevotella copri*, exist in the guts of Chinese and Americans. Other important commensal species commonly found in the healthy gut, such as *Faecalibacterium prausnitzii*, also vary across global populations because of subtle genetic variance in the microbial code.[17] This is important, because bacteria like *F. prausnitzii* are highly active members of the microbiome, influencing lots of human pathways of metabolism; their relative presence or absence might explain variations in disease risk across populations. Only a very few strains of bacteria occur in multiple unrelated cohorts; the *Bacteroides* species appear remarkably consistent, whereas *P. copri* is among the most variable gut colonizers.

We're only scratching the surface of understanding the profound impact that Westernized living has had on our internal ecology. Palaeofaecal samples from our ancestors are more similar to the gut microbiomes of modern non-industrialized humans than to those of industrialized humans, and the modern gut harbours far higher abundances of antibiotic resistance and mucin-degrading genes than the ancient gut.[18] Changes in the ecology of the gut map directly on to the burden of chronic disease: colon cancer and inflammatory conditions of the bowel have an extremely high prevalence in the West and very low rates in rural Africa.

A small number of studies of indigenous peoples who still live traditional lifestyles have started to examine the major environmental forces that are driving this variance.[19] These have included communities from Venezuela, Peru, Tanzania,[20] Botswana,[21] Malawi and Madagascar.[22] But the vast majority of our research comes from the US, Europe and China and substantial gaps remain in our description of the microbiome across global populations. For instance, of the sixty-four studies surveying the gut microbiome of individuals living within Africa as of January 2021, only twenty-five of the fifty-four countries on the continent were represented.[23] Our knowledge doesn't reflect the true global state of the human microbiome, and we have much to learn from the dynamic movement of people across the globe, and between rural and urban environments.

Urbanization

Today, 55 per cent of the world's population lives in urban areas, and this will rise to 68 per cent by 2050. The UN projects that urbanization, combined with the overall growth of the world's population, could add another 2.5 billion people to urban areas by 2050, with close to 90 per cent of this increase taking place in Asia and Africa.[24] Tokyo is the most populated city on the planet, with more than 37 million people living within its metropolitan area; according to the International Health Metrics Evaluation, the disease that kills the most people in this highly developed, digital city is Alzheimer's, and four types of cancer also make the top-ten-killers list (lung, colorectal, stomach and pancreatic). However, the city with the highest population density in the world is Manila in the Philippines, which hosts 119,600 people for every square mile. There is a thirteen-year difference in life expectancy for residents in the two cities, with those living in Manila facing much greater risks of diseases associated with deprivation, such as ischaemic heart disease and stroke, as well as TB and interpersonal violence. This is a tale of two cities on two different GEM lines.

Urban humans spend disproportionately more time indoors than rural dwellers, and the architectural design of our buildings influences our exposure to organisms. In fact each building we live in has its own microbial biogeography. Scientists from the University of Puerto Rico found that microbes from house walls and floors segregate by location, and urban indoor walls contain human bacterial markers of space use.[25] The design of the buildings we live and work in influences not only our risk of exposure to pathogens, but also our chances of being exposed to symbiotic microbes that could be beneficial. This might be key to understanding our susceptibility to developing allergies. For example, house-dust microbiota are significantly different in rural and urban areas, with higher relative abundances of Ruminococcaceae, Lachnospiraceae and Bacteroidaceae families in rural houses compared with urban houses. Children under five exposed to higher indoor concentrations of the dust microbiome also experience an increased risk of lower respiratory-tract infections.

The gut microbiomes of city dwellers have more metabolic pathways to degrade drugs and organic compounds, and they have lower abundances of rural *Prevotella* spp. and fermentative *Clostridiales*, with Enterobacteriaceae and *Bacteroides* spp. useful for metabolizing a diverse diet. But the design of our cities also influences the bacteria that we are exposed to. In 2021, a global atlas of 4,728 metagenomic samples from mass-transit systems in sixty cities over three years was published, representing the first systematic, worldwide catalogue of the urban microbial ecosystem.[26] This stunning piece of work identified 4,246 known species of urban microorganisms, and a very small number of these species (just thirty-one in fact) were distinctly different from human commensal organisms found in the gut and on the skin. This implies that cities possess a small but consistent core set of non-human microbes that may be important for our health. Exactly like East and West Berlin, each city hosts its own unique set of geographical, cultural and environmental niches.

It's likely that air pollution is also contributing to changes in the composition and function of the human gut microbiome. Long-term exposure to micro-scale particulate matter, nitrogen dioxide (NO_2), ozone (O_3) and nitrogen oxides (NOx) is associated with a greater risk for obesity, glucose dysregulation and type II diabetes. We breathe in ultra-fine pollutants and these make their way into our blood circulation and to our gut. A study in California demonstrated that exposure to ozone for twenty-four hours was associated with reduced gut microbial diversity and increased abundances of *Bacteroides caecimuris*. Nitrogen-dioxide exposure was also associated with changes in gut ecology with the loss of a number of taxa. Before you've even had the chance to dive into a plastic-contaminated sea, the air you breathe has already changed your gut microbiome.

Indigenous people moving in and out of rural and urban environments change their microbiome and its functions as it adapts to these environmental stresses and strains. But there is more to this story than environmental contaminants.

Modern family

Families tend to share the same environment and the same exposome, and just as social interactions shape wild primate microbiomes, the same is true for humans. Family or kin relationships characterize all societies because they allow a safe and supportive social structure for reproduction, socializing, education and economic stability. There's an evolutionary benefit to maintaining robust microbiomes within family networks.

The average family size varies significantly across the globe; if you live in Senegal the chances are that you will be living with 8.33 other people, and I can only imagine how complicated breakfast is. If you live in Germany there will be 1.91 other family members in your life, and it will be much quieter. Globalization is changing our family size and dynamic and our social networks, which in turn modifies our ability to exchange bacteria.

Within typical families, the structure of the microbiome is pretty stable and highly specific to family members. Scientists at the University of Chicago followed microbial communities associated with seven families and their homes over six weeks.[27] This cohort included three families who moved their home during the study period. They found that microbial communities differ substantially among homes, and the home microbiome is predominantly seeded and sourced by the humans who inhabit it. These observations were so strong that it was possible not only to identify family homes by the house microbiome, but also to match individuals to their houses through their respective microbiomes. Even potential human pathogens observed on kitchen counters could be matched to the hands of occupants. Most interestingly of all, the microbial community of the occupants' former house rapidly converged on the microbial community in a new house. So when you move home, you literally move your bugs with you as part of the decoration strategy.

How often you interact with your family will also play a role in defining the bugs you share. For example, going home for Thanksgiving or religious holidays is an important method of reconnecting with the family microbiome. Researchers performed a metataxanomic analysis of faecal samples collected on 23 and 27 December from a group of twenty-eight healthy volunteers celebrating Christmas.[28] They found two distinct microbial-biomarker signatures: those who had decided to see only their immediate family over the Christmas season, and those who'd visited both their own family and their in-laws. Participants visiting in-laws demonstrated a decrease in all *Ruminococcus* species, which may mean they were not being made to eat Brussels sprouts. This type of bacteria has also been associated with psychological stress and depression . . . Make of that what you will. But the sharing of microbiomes between family members may not always be a positive experience.

Several attempts have now been made to study the dynamics of family microbiome networks. The Fiji Community Microbiome Project (FijiCOMP) explored the role of bacterial transmission in human populations in 287 people living in five agrarian villages in the Fiji Islands.[29] Within this small community of individuals with homogeneous living environments, diets and microbiomes, bacterial DNA alone could be used to accurately predict certain intimately linked pairs of individuals. Strong transmission patterns were also found within households. Women were more likely than men to harbour strains that were closely related to those found in family members and social contacts.

The implication of this work is that our relationship with microbes is not only determined by the buildings we live in, but also by the people we share them with. It also means that the susceptibilities some families have for specific chronic diseases might be driven, in part, by the presence of microbes that come from the people we live with. Further evidence of this has come from twin studies. The gut microbiome of twins where one has inflammatory bowel disease (IBD) and the other doesn't demonstrates that the healthy twin displays IBD-like microbiome signatures, even though they have not developed the disease.[30] We don't currently know if the IBD-like microbiome signatures precede the onset of IBD, but it's plausible that the environmental microbiomes of our families determine part of the risk of this disease, as well as the risk of not getting the disease, as we saw in the allergy twin study in Chapter 4 (see page 63).

Modern families are increasingly likely to be geographically dispersed, and we travel further to be with our loved ones. The aeroplanes that churn out fossil fuels into the upper reaches of our atmosphere don't simply burn holes in our ozone layer; they also transfer pathogens, symbionts and amensalistic microbes in their human passengers across a vast web of air corridors. Container ships, mechanized transport and the internet carry our diets, livestock, behaviours, knowledge, conspiracy theories and social norms around our cities and to every corner of the globe and even into space. The physical and metaphysical mobilization of humankind is fundamentally redrawing the infectious-diseases map. It allows pathogens to move quickly; COVID-19 had spread across the globe by the time the first patient in Wuhan was ventilated. And as the world gets warmer, the mosquitoes that carry malaria are using this network to move away from the Equator into parts of northern Europe where they can now live quite happily. Higher global temperatures driven by climate change are selecting for microbes with thermal tolerance that have pathogenic potential for mammals. This is what happened to the dinosaurs.

The transport of holobionts has also had a great impact on the spread of chronic disease. In 2020, the International Organization for Migration estimated that there were 272 million migrants in the world, accounting for 3.5 per cent of the world's population.[31] Many of these migrants will have moved for work, but millions have been driven by conflict, violence and climate change. Migration on this scale brings with it multiple social and health challenges. This doesn't mean that migrants are disease-carriers who present a risk to public health or that they're a burden. On the contrary, migration typically benefits national and global economies, and is more harmful to the migrant in terms of their individual risk of non-communicable disease. Migrant studies, such as those performed in Japanese Hawaiians, have demonstrated that it only takes one generation before the immigrant population assumes the risk of chronic diseases of the West, such as colon cancer.[32] Although a change in diet plays a part in this risk, migration alters many other aspects of the exposome. For example, cigarettes, chemicals, infections and antibiotics might be equally responsible for modifying the microbiome and, in turn, non-communicable disease risk.

In March 2020 we suddenly stopped flying, and we locked up our offices, our schools and our elderly. In an instant we shut down the circulation of microbes throughout our global and local communities, and we replaced these intricate networks with Zoom calls and loneliness. The true impact of this change on the global microbiome is not known, but just when things had got back to normal, Vladimir Putin started a war that sent oil prices sky-high, changing once more how we travel, cook, warm our homes and pollute our environment. Our microbiomes will never be allowed to rest.

Tortola

The British Virgin Islands have had a turbulent history; their population has endured slavery, cholera epidemics and endless invasions by Europeans. Today, they must deal with urbanization, Americanized diets and climate change. The total population of Tortola grew by approximately 15 per cent between 2010 and 2022, with more than 60 per cent of the island's population now made up of expatriates, many of whom work within the booming financial-services industry. They are bringing with them their urban microbiomes to fill their new homes, and they are changing the social exchange of bacteria on a small island. Tortola is still highly dependent on its stunning landscape, beaches and 120,236,505m^2 of fragile coral reefs for its economic and social well-being. Holobionts no longer come by rickety aeroplanes, but by gargantuan cruise ships and charter yachts, full of pathogens, petrol, plastic and money. In 2019, 894,991 people visited the British Virgin Islands. I'm not sure what my grandmother would have made of it. I suspect she would have lit another cigarette.

My hope is that recognizing the immediate climate emergency happening inside all of us will help motivate us to make the right choices. Saving the planetary and individual microbiome from hyperglobalization doesn't only mean consuming less, recycling more, planting trees or driving electric cars. It requires us to completely reappraise our relationship with microbes and to prioritize them. Our cities, urban spaces, offices, hospitals, schools and homes need to be redesigned to promote microbiome diversity, and our planetary microbial systems must be sustained as part of our climate-change policy. Our health can't be sustained by the metaverse – it demands real-world connections.

Our lust for manufacturing is forcing microbial life forms into damaging homogeneous niches, and their belligerent behaviour simply reflects the fact that they're also struggling to cope. To survive they have had to adapt to eating plastic, petrol and pesticides, and I can't believe those taste that great. If we offer them a little bit of love, these microbes can be leveraged for sustainable farming and engineered to recycle CO_2, carbon and nitrogen, to produce oxygen and even to remove the toxic waste polluting our environments and our bodies. For this, we have to be prepared to innovate, and we have to be prepared to pay for it. A truly global response means that we have to empower all nations and cultures to analyse and understand their own microbiomes, and we can't rely on scientific organizations suffering from colonial hangovers to selectively hand down knowledge of such importance.

The fourth age of man needs the microbiome.

10. The War on Bugs

Gentlemen, you can't fight in here! This is the War Room!

Dr Strangelove

In 1493, Columbus departed from Cádiz for his second trip to the Americas. This time he was planning to set up shop and so he took with him 1,000 colonizers in seventeen ships. In a letter to King Ferdinand and Queen Isabella of Spain, he reported:

> I brought with me horses, mares, mules, and all of the other beasts, as well as seeds of wheat and barley and all the trees and all the choice fruits, everything in great abundance.

He also brought with him trillions and trillions of microorganisms. The party landed in the Americas on 8 December, and as soon as the hogs were released, there was an outbreak of swine flu that swept through the crew and the indigenous population. This was the first great American pandemic, and it was malignant and terrible. It was also the beginning of the 'Columbian exchange', and the transfer of diseases, ideas, foods, crops, tobacco and animals between populations that had been kept apart for 12,000 years. It marked a dramatic turning point in the phylosymbiotic evolution of the human microbiome. It is hard to conceive that the humble tomato or the ubiquitous potato did not exist in Europe until this moment. Today, in the US these account for almost half of all vegetables consumed. The conquistadors' guts must have bloomed with happy bacteria, merrily munching on the fibres and polyphenols of the New World. Yes, they would also have had to deal with the occasional tropical amoeba, helminth, nematode or parasite, but this was nothing compared to the indigenous populations, who succumbed to terrible European pathogens such as smallpox, for which they had no immune defence.

Five years later, on 20 May 1498, the Portuguese explorer Vasco da Gama became the first European to reach India via the southernmost tip of Africa. His mission was to find a cheap and plentiful source of spices, and his arrival on India's Malabar Coast, in modern-day Kerala, marked the start of direct trading between Europe and South-East Asia. His discovery meant that the Portuguese controlled the coastal spice-producing region and had a wildly profitable monopoly on the Indian spice trade. This in turn meant global conflict. What followed was the worst of colonialism, with all the European superpowers greedily vying for control of global trade routes. The ultimate victor would be the British East India Company, a commercial institution founded in 1601. Over multiple spice wars, it would overtake the Dutch and become one of the most powerful and enduring private organizations the world has ever seen. The Company collected taxes, trafficked slaves and committed atrocities in the name of the queen who had given the company a royal charter, enabling it to wage war. It seeded the growth of the British Empire until 1858, when the British government finally ended Company rule in India. In just over 400 years, the human microbiome was completely transformed through the creation of a global trading network that also spread knowledge, symbionts and pathogens. These things were not shared equally.

All empires require battles to be fought by soldiers, and the co-ordinated spilling of blood has had a significant impact on microbial life forms. Today, the world spends US $1,981 billion on defence, and humans show no sign of losing enthusiasm for kinetic – and increasingly autonomous – killing. Governments have weaponized pathogens and created murderous biological warfare programmes designed to trigger pandemics and to poison and suffocate entire cities. More recent wars have placed an intolerable pressure on the global microbiome through environmental damage, the industrialization of food production and the displacement of millions of refugees, who often have nothing but their microbiomes to their name. Because of the inherent variation in the environment, region, tribe or choice of weapon, each conflict also has its own microbiome, and the consequences of its destruction are long-lasting and multi-generational, for combatants and innocent bystanders alike. To understand how and why, we need to take a historical perspective on conflict and its broader environmental and ecological impact.

Killing in the name of . . .

Modern militaries have served as a major driver of antiseptic innovation, and their love of expeditionary warfare and logistics has meant they've successfully scaled the use of antibiotic and pathogen prevention strategies across the globe. But before germ theory, military surgeons stood in fields with bare hands and butchers' aprons and poured wine or boiling oil into gunshot wounds. Amputation often served as a form of palliation, and the best measure of surgeons' skill was their speed. In the US Civil War, amputation was the most commonly performed surgical procedure for the 60,266 Union patients who sustained gunshot fractures, and gangrene had a 60 per cent mortality rate.[1]

As soon as Louis Pasteur fired the starting gun on germ theory, war became as much about killing bacteria as it was about killing the enemy. The well-described innovations in hygiene and antisepsis developed by Florence Nightingale, Ignac Philipp Semmelweis and Joseph Lister provided military physicians with simple tools through which they could limit the fatal contamination of wounds and treat sepsis. I roll up my sleeves and wash my hands every time I examine a patient (and we all covered our hands in alcohol during the coronavirus pandemic) as a direct result of these visionaries. Even though Lister would liberally wash everything in carbolic acid, he apparently did not believe in sanitation, and he would operate by removing his jacket before pinning an unsterilized towel over his waistcoat. I can only assume he must have spent a fortune on dry cleaning.

On more contemporary battlefields, the number of casualties

dying of wounds has gradually declined from 8 per cent in the First World War to 4.5 per cent in the Second World War, 2.5 per cent in Korea, 3.6 per cent in Vietnam and 2.1 per cent in Desert Storm.[2] The widespread adoption of antibiotics also made a major contribution to these outcomes. Today, an army marches on its vaccines, although these are not compulsory. During the 2003 Iraq War, for example, the British military decided to emphasize the voluntary nature of the anthrax vaccination.

Between 1941 and 1945 the medical corps urgently needed a prevention strategy for the rising cases of communicable diseases such as strep throat.[3] By D-Day their chosen chemical antibiotic, called 'Prontosil', had to be abandoned entirely. In just five years the industrialized antibiotic programme had caused bacterial resistance – a new strategy was needed on the front line. It had become clear to the British scientists developing penicillin at the start of the war that they didn't have the industrial capacity to scale up its production and, after a lengthy negotiation, the US pharmaceutical giants Merck, Squibb, Pfizer and Abbott took over its production. British and American production grew from twenty-one billion units in 1943 to 6.8 trillion units in 1945.[4] In November 1942, penicillin was first administered to US troops wounded during an assault in Oran, Africa, although the first really large-scale military use was during the D-Day invasion of Normandy in June 1944. Penicillin did not win the war, but one estimate suggests it saved the lives of about one in seven soldiers wounded in battle, as well as permitting more invasive lifesaving treatments to be performed. The human microbiome would never be the same again.

Biopolitics shape the microbiome

By 1933 the French-Canadian self-taught microbiology maverick Félix d'Hérelle had founded the Laboratoire du bactériophage for the development and production of phage therapy. His work had originally been funded by the British-Indian government, which had in 1927 opened the 'Bacteriophage Inquiry' to evaluate phage properties in the treatment and prophylaxis of cholera and plague, with amazingly successful results.[5] Giorgi Eliava, a Georgian physician and bacteriologist who had worked with d'Hérelle during an extended visit to the Pasteur Institute in 1918, became another key proponent of bacteriophage research. Eliava and the left-leaning d'Hérelle became firm friends, and in 1923 Eliava founded a bacteriological research centre in Tbilisi with the blessing of Soviet dictator Josef Stalin, who funded much of this work. In 1933, d'Hérelle left Yale University to join his protégé, and between 1934 and 1936 he established two more phage production laboratories in Kiev and Kharkov, by government invitation. D'Hérelle stayed in Russia until 1937, when Eliava and his wife were arrested for the crime of being intellectuals. It is thought his actual crime was to fall in love with the same woman as Lavrentiy Beria, chief of the secret police to Joseph Stalin. Whatever the reason, Beria deemed it serious enough to have both Eliava and his wife executed. Despite Eliava's death, the institute continued its pioneering work, and in the 1940s it developed phage against anaerobic infections such as gangrene. The Red Army physicians used phage therapy during the war in Finland, and it is alleged that German troops expressly occupied Georgia in the Soviet Union to seize the phage produced at one of d'Hérelle's scientific centres.

Meanwhile the Americans were not exactly twiddling their thumbs on phage. The famous 'phage group', an informal network of scientists begun in 1940, was founded by the physicist Max Delbrück and also made major insights into this area of biology. Jim Watson, who went on to discover the double-helix structure of DNA, performed his PhD under Delbrück's supervision and demonstrated that phage can participate in genetic recombination. However, the Allied health strategy during the Second World War was firmly based on antibiotic therapy.

The Russian military's love of phage did not slow after the Soviet Union broke up. In the 1990s, Georgian soldiers fighting in the breakaway Abkhazia region carried spray cans filled with phage against five bacteria: *Staphylococcus aureus*, *Escherichia coli*, *Pseudomonas aeruginosa*, *Streptococcus pyogenes* and *Proteus vulgaris*, and phage continue to be widely available in many Russian cities, where they can be bought in a pharmacy without prescription. Phage therapy and prophylactic measures became ideological symbols of divisions and disagreements between Western and Eastern countries. In the West, antibiotics would become the weapon of choice – and their misuse would have serious implications for human ecology.

At the end of the Second World War the need for effective antibiotic therapy continued and became a source of continued tension between East and West. Penicillin was of such immense importance that it was used as an espionage tool at the start of the Cold War. As part of Operation Claptrap, the US intelligence officer Major Peter Chambers attempted to bribe Russian soldiers to share Red Army secrets in return for penicillin to treat their gonorrhoea and syphilis.[6] During the 1940s the Soviet Union and China had developed some penicillin production capabilities, and this production initially increased after 1945 when the Allies shared their knowhow. But during the Cold War antibiotic production became a source of competition. By the 1950s the USSR and Eastern European countries had found a way to produce generic antibiotics, based on stolen Western antibiotic technology. The war on bugs was now fully industrialized, and it was global.

Resistance is futile

The humanitarian response to recent conflicts in the Middle East has been dominated by their local environmental microbes and pathogens. After the Allied invasion of Iraq in 2003, US military surgeons began reporting on a multi-drug-resistant bacterium they called 'Iraqibacter'.[7] The actual organism was *Acinetobacter baumannii*, a bacterium that only served to increase the true horror of wounds because they became infected or didn't heal. This was far from the only resistant organism. In Afghanistan, Allied forces had to deal with high rates of aggressive multi-drug-resistant bacteria like Panton-Valentine leukocidin (PVL) producing *Staphylococcus aureus*,[8] which caused terrible soft-tissue infections. As if war were not bad enough, modern-combat medics must now deal with antibiotic stewardship and multi-drug resistance.

But combat doesn't only influence our relationship with pathogens – it also mutates our commensal and amensalist microbes, and this is just as important.

Gulf War syndrome is a chronic and multi-symptomatic disorder affecting military veterans of the 1990–91 Persian Gulf War. A wide range of acute and chronic symptoms have been linked to it, including fatigue, muscle pain, cognitive decline, insomnia, rashes and diarrhoea – all symptoms that are typically seen when the immune system isn't functioning. While the cause of the condition is debated, it's thought to be related to exposure to pills containing pyridostigmine bromide, a pre-treatment for nerve agents, pesticides or chemicals from oil wells, among other compounds. These agents not only perturb the gut microbiome, but also injure the gut barrier, causing a systemic inflammatory response in animal models.

A proportion of veteran soldiers suffering from this condition have gastrointestinal symptoms, and the type of microbes found in their gut varies from those who don't have symptoms. Specifically, they have greater abundances of Bacteroidetes, Actinobacteria, Euryarchaeota and Proteobacteria, as well as higher abundances of the families Bacteroidaceae, Erysipelotrichaceae and Bifidobacteriaceae. These bacteria are linked to increased levels of inflammation in the blood and the worsening of other symptoms.[9] This could, of course, be an effect rather than a cause, but the microbiome is implicated.

Of the 15,000 veterans surveyed in one study, nearly 50 per cent were overweight and about 30 per cent were obese. Obese veterans were more prone to developing other chronic health conditions such as post-traumatic stress disorder (PTSD).[10] The gut microbiome is intimately linked to obesity and mental health through the gut–brain axis, and it's plausible that it is playing an indirect role in causing the symptoms of Gulf War syndrome. A series of mechanistic events might look like this: an environmental trigger during deployment in a war zone perturbs the gut microbiome of a susceptible individual, and the microbiome is then unable to recover to its pre-deployment structure or function. This disrupts the immune system and the gut–brain axis, which causes fatigue and exacerbates the risk of obesity and gut dysfunction. It is also possible that this shares a common mechanism with conditions such as myalgic encephalomyelitis (ME) and even Long Covid, where sudden manipulation of the immune system causes non-specific, systemic symptoms that cannot easily be treated. This is just a theory, although war most definitely shapes both human and planetary microbiomes through very large and destructive environmental forces.

Climate conflict

Over the course of human history, wars have served as a form of local and global population control. Genghis Khan was a prolific and highly efficient killer of men. He started out by killing his brother, and by the end of his reign was responsible for the death of 11 per cent of the global population. He also weaponized the horse, ensuring that the zoonotic transmission of microbes would become a fixture of all future conflicts. However, the Second World War comfortably wins the prize for the most casualties in modern times, with sixty-six million dead.

The fact that ancient Mongol fighters achieved casualty rates comparable to those of modern mechanized combat has nothing to do with swords or gunpowder. In the conflicts of old, the real harm was caused by plagues and pathogenic microbes, which decimated armies, civilian populations and refugees. War is an astonishingly effective method of spreading pathogens and of damaging the microbiome of indigenous people through famine, displacement, toxins and inflammation. It is also an effective method for reshaping our terrestrial and marine ecosystems, which in turn regulate our own.

The consequence of the biological warfare inflicted by the conquistadors on the local environment and on the internal microbiology of the indigenous peoples was the apocalypse. By 1550 approximately twenty-four million Aztecs were killed, with 1.5 million in the Mayan states and eight million Incas. Influenza, smallpox, measles and mumps killed millions of these people, and researchers now believe that 'cocolitzli' – a typhoid-like enteric fever – was largely responsible for the majority of deaths that finally led to the collapse of the Aztec empire. Researchers at University College London (UCL) have concluded that this loss of life resulted in almost fifty-five million hectares of agricultural land being untended, an area the size of France.[11] In time, the forests returned to their native flora and fauna and they began to absorb CO_2 and produce oxygen. The regrowth was so significant that it soaked up enough carbon dioxide to cool the planet, with the average temperature dropping by 0.15°C in the late 1500s and early 1600s. This was enough that in 1608 the River Thames froze over in London.

Just as the global temperature fell after the Aztec apocalypse, so the world's temperature also fell after the Second World War. But this happened for a quite different reason. Despite the carnage, humans had managed to continue to reproduce, and the global population rose from two billion before the war to approximately 2.3 billion. This time there were no empty farms; instead, the industrial economies of Europe and the US achieved a level of productivity the world had not previously seen. Expansion on this scale meant that an enormous number of fossil fuels (coal, oil and natural gas) were burned. This did not only produce carbon dioxide, but also produced particulate matter, including soot and light-coloured sulphate aerosols that kept the sunlight out and cooled the planet. These changes were short-lived, and the greenhouse effect would soon come to dominate, dramatically warming the planet, as it does to this day. The conflict of the Second World War shaped our microbiome; however, it was the industrialization of our farming practices and the massive growth in manufacturing that fed our addiction to a globalized diet, and our dependence on pharmaceuticals and antibiotics that did the real damage to our internal ecology.

In the last seventy years we have created incredibly efficient war machines that are designed to destroy the natural world. Chemical agents such as mustard gas have numerous toxic effects on soil microorganisms and inhibit their enzyme activity. High-explosive munitions leave heavy metals in the soil that exacerbate antibiotic resistance, and anti-personnel mines can render land unusable and harm wildlife. Anti-plant agents like Agent Orange, employed by the US in South Vietnam, decimated precious jungle ecosystems. In 1991, 800 oil wells were destroyed in Iraq, and twenty million tonnes of leaking oil formed networks of black rivers covering 200 km² of ground under a dark and toxic cloud of burning oil.[12] War and the preparation for war create a 'conflictome' of perturbed environmental and human microbes that cause harm long after the shooting has finished.

Climate change exacerbates the displacement of people caused by war; the UN estimates that 89.3 million people worldwide were forcibly displaced across borders through conflict in 2021.[13] You can take your pick from a number of refugee camps to tell this story: Kakuma in Kenya, Za'atari in Jordan or Um Rakuba in Sudan. All host vulnerable people at the mercy of pathogens and disease vectors, and many of these camps have been devastated by COVID-19. These dusty, crowded and noisy makeshift cities, which cradle delicate human lives, serve as a perfect environment for horizontal gene transfer. Metataxonomic analysis of 500 refugees, from Syria, Iraq and Afghanistan to Germany, showed that the new arrivals had a distinct microbiome from that of their European hosts, with more Bacteroidetes and fewer Firmicutes, which are most likely representative of poor nutrition, and with a higher burden of antibiotic-resistant genes.[14] Children and families subjected to the torment of conflict also suffer from the chronic diseases and mental-health problems you would expect of a traumatized and deprived population. The loss of microbial symbionts might play an important part in determining their resilience against pathogens and chronic disease, and the microbiome represents a novel target for protecting these vulnerable populations.

The home front

Starvation is an inevitable consequence of any conflict and, after antibiotics, food is one of the most important modifiers of the gut microbiome. In March 1945, the British War Cabinet had to deal with a nation facing serious food shortages and a rationing crisis. With a prospective deficit of 1,800,000 tons of meat, something had to be done.[15] Farmers were encouraged to adopt the newly discovered antibiotic prophylactic strategies to prevent disease, reduce costs, maintain production, preserve food and minimize economic risk. Pullorum disease in poultry, for example, is caused by the bacterium *Salmonella pullorum*. It is vertically transmitted and typically affects young chicks, and it has a high mortality rate. Before the war, farmers had to resort to scrubbing hatching trays and eggs with hot lye, or using carbolic acid or chlorinated lime. Eggs and hatching chicks were even fumigated with oncogenic and highly toxic products, such as formaldehyde or potassium permanganate.[16] In 1948, Merck's sulfaquinoxaline became the first antibiotic to be officially licensed for routine inclusion in poultry feeds against the protozoa coccidiosis. The miserable life of the broiler chicken was extended and the 'chickenization' of our Western diet began. It was not finger-licking good. Today, in the UK, 95 per cent of these sorry animals are farmed industrially to meet our consumption of three million chickens each day.

The global uptake of antibiotics in farming hasn't solely been driven by the need to prevent diseases of overcrowding and unsanitary feedlot conditions. As early as 1946 scientists observed that antibiotics triggered more efficient growth in poultry and swine when they were fed the mycelia of the fungi *Streptomyces aureofaciens* that contained chlortetracycline antibiotic.[17] The consequence of this was dramatic. In 1935, it took 112 days for an average chicken to reach the market weight of 1.27 kg. By 1985 it took just thirty-five days to get to 1.4 kg, and today over the same time it weighs a whopping 2.44 kg.[18] This has been caused by a combination of interventions, but feed antibiotics can increase the body-weight gain of a chicken by up to 8 per cent and decrease the feed conversion ratio (feed intake: body-weight gain) by up to 5 per cent when compared with an antibiotic-free diet. This was good news for farmers as antibiotics were cheap. The effect was so dramatic in all forms of livestock that a boom in antibiotic growth products began, and the US Food and

Drug Administration (FDA) approved the use of antibiotics as animal additives without veterinary prescription in 1951. By 1958 up to 50 per cent of British pigs were fed antibiotics, and nearly all unweaned piglets had access to food containing tetracyclines. In 1966, West Germany's Minister of Agriculture estimated that 80 per cent of mixed feeds for young pigs, veal calves and poultry contained antibiotic additives.[19] It's very likely, then, that antibiotics were a key reason for the microbiome dysfunction of West Germans that resulted in their higher rates of immune-mediated diseases compared to East Germans, during the Cold War.

But it wasn't only the West that was ramping up meat production. In East Germany the GDR built industrial high-rise pig and poultry farms that were far worse than George Orwell could ever have imagined. Known as the *Schweinehochhaus*, these high-rise monstrosities were supposed to be a symbol of communist ingenuity. More than 500 sows could be grown over several floors, carried up and down by lifts . . . They also needed a lot of antibiotics. Since then, meat consumption has continued to rise spectacularly all over the world. China now consumes 28 per cent of the world's meat, including half of all pork produced. Brazil supplied 43 per cent of China's meat imports in 2020, and the majority of these cows were reared in a vast tropical region of the Amazon called the Cerrado; half of the Cerrado and about 20 per cent of the Brazilian Amazon have now been cleared to facilitate this growth.

During the Second World War the world's milk larder ran dry. More cows were needed, and they in turn needed more land mass and feed to support their growth. Each cow had to be efficiently milked so that it could reliably provide its 2,725 litres per year.[20] Today this figure has swollen to well over 10,000 litres per year. In 1940, the antibiotic Gramicidin was first used to treat an outbreak of mastitis in cows from New York, and by 1943 precious penicillin supplies were tested against mastitis in both Britain and Denmark, and milking tubes treated with antibiotics against mastitis proved wildly popular. By the 1950s up to 10 per cent of US milk samples were contaminated with penicillin, caused by overdosing of animals or the illegal use of antibiotics.[21]

Synthetic antibiotics had entered our food and our drink, and we have a world war to thank for it. In the UK today there are more than one billion animals being farmed for our use, dwarfing the human population of just sixty-five million people. Methanogens in the intestines of these animals produce so much methane that meat farming now accounts for 60 per cent of all greenhouse gases emitted by the food-production.

Farming, aquaculture and processing plants became huge consumers of antibiotic arsenicals, sulphonamides and quaternary ammonium compounds; by the 1960s, these chemicals entered our food supply. And this wasn't only taking place on land. From 1970 onwards, salmon farming boomed along the Scandinavian North Sea coast, and bacterial and fungal infections fostered the routine use of antibiotics like oxytetracycline, amoxicillin and sulfadimethoxine-ormetoprim. Almost 80 per cent of these antibiotics entered maritime environments, where they were selected for antibiotic resistance in sediments, wild fish and shellfish populations. Today, more than half of the world's human population gets 20 per cent of their protein from fin-fish and shellfish; in 2018, at least sixty-seven antibiotic compounds were used in aquaculture to sustain this growth in eleven of the fifteen major producing countries, the largest of which is also China.[22]

Despite the industrialization of our post-war farming practices, food security and famine during conflict remain a significant threat to life, and this burden is disproportionately borne by countries in Africa. In South Sudan, more than one million people are food insecure. In 2018, the UN adopted Resolution 2417 and formally condemned starvation as a weapon of warfare, yet today we stand on the precipice of a global food catastrophe, once again precipitated by war in Europe.

Biosecurity and the military microbiome

A raft of conspiracy theories currently exists about the SARS-CoV-2 virus origins in a government facility in Wuhan. While there is no convincing proof yet for this accusation, there are more than 3,000 biosecure laboratories worldwide that we know of, and many governmental agencies are courting biohackers to join the fight in the rapidly growing field of biosecurity. The future of fighting is going to include the use of synthetically engineered microbes as offensive weapons. The awful truth is that, with the global access to CRISPR technologies (see page 16), it has never been easier or cheaper to develop pathogens to cause harm. Just as smallpox was weaponized by the conquistadors, so similar organisms can be developed for either state-on-state or terror attacks. These frightening autonomous microbes will not only be programmed for disability, lethality or plague, but also for information-gathering, cryptography or even target identification.

Because of COVID-19, organizations such as the UK's Joint Biosecurity Centre (JBC) now provide central information on the spread of pathogens and the risk of emerging biological threats. But I find it interesting that almost no government agency focuses on the bioresilience of our microbiome as a fundamental form of national defence. An optimal microbiome doesn't mean you're not going to get COVID-19, but it might reduce your chance of dying from it. The microbiome is a major opportunity for ecological engineering against offensive pathogens. The beauty of it is that we don't need fancy synthetic biology, just a government that really wants to promote gut health. And this strategy will not just be of benefit to humans. The World Organization for Animal Health (WOAH) reported twenty-two million cases of bird flu in wild birds across sixty-eight countries in 2021–2, leading to the killing of 94.2 million birds.[23] Our farm animals could do with some bio-resilience too.

Military medics of the future will move beyond antisepsis towards the adoption of microbiome theory. They'll do it not only to maintain the fitness of their fighting force operating in a range of hostile environments, but also to minimize the consequences of physical trauma when the worst inevitably happens. A strategy of 'supra-organism trauma care' optimizes critical microbial functions while using a precision strike on any bad actors, rather than the nuclear option of eradicating every single bacterium with broad-spectrum antibiotics, which will only cause more malignant strains of 'Iraqibacter' to emerge. The microbiome influences everything from wound-healing to blood clotting to organ regeneration, and it might even influence our susceptibility to PTSD.[24] Looking after it when you are hurt is a good idea and, like so many other trauma innovations, this will eventually trickle down into civilian care.

Through conflict we destroy the mutualistic symbiotic relationships that we depend on for our health. War changes our economy, culture, climate and behaviour, and it displaces people – and so wars change our microbiomes in ways beyond antibiotic consumption and drug use. Conflict is a major evolutionary force that shifts the entire GEM line for generations, and to date none has done more to influence our microbiome than the Second World War. In the twenty-first century global conflict sadly remains a regular occurrence, and multiple generations have now experienced 'conflictomes' that have disrupted the orchestral signalling that is passed from mother to baby. It comes as no surprise, then, that these traumatic and upsetting experiences cause chronic diseases that emerge only after the battle finishes, such as cancer, Gulf War syndrome and PTSD.

Centuries of increasingly global conflicts have placed unimaginable stress on the human mind and our environment, and the microbiome is responding by doing what any human population would do if attacked: it has gone to war. The global antimicrobial strategy was accelerated by war and neutered by the failure of biopolitics, and it is depressing that history is once again repeating itself. Foreign policy exerted through military force tends to fail, and our foreign policy on microbes is no different. Over the last few hundred years we have carpet-bombed the human microbiome for minor infringements on our health, and its population of innocent symbionts has been destroyed; it now represents a cratered field in Flanders, without trees or birdsong. The result is that we are all suffering from a microbiome stress disorder. The solution is a peace treaty, but for now I can't see anyone with a white flag and a pen.

11. A Biotic Life

When my daughter was four months old she developed a cough and a high-pitched wheeze. Her temperature climbed and she stopped feeding. The Girl was worried, but I knew best. It was bronchiolitis, caused by the common Respiratory Syncytial Virus (RSV), and it would most likely get better. We gave her paracetamol (Tylenol), but she didn't improve and the Girl said we should go to hospital, but I still knew best. Twenty-four hours later and Liberty Kinross was deathly pale, dehydrated and exhausted. I did not know best.

Evidence from a retrospective analysis of 13,516 births in Israel, where one sibling was delivered vaginally and one was born by Caesarean section, suggests that the child born by Caesarean is more than twice as likely to develop bronchiolitis within the first two years of life.[1] Liberty was duly admitted to the paediatric high-dependency unit at London's St Mary's Hospital in Paddington, where she was diagnosed with bronchiolitis and a superimposed bacterial pneumonia. She was resuscitated with oxygen under a warm lamp. A white plastic tube was passed through her nose and into her gut, so that she could be fed with infant formula. Intravenous lines were placed into her chubby arms, so that she could be given the antibiotics that would save her life. As I sat in the dark by her cot, I contemplated my stupidity and I wrestled with my guilt. Any parent who has kept watch by the bed of their seriously ill child will know how exquisitely painful this is.

The place where she was being cared for is important in this story. It was at St Mary's Hospital that, in 1928, penicillin was first discovered by Sir Alexander Fleming. His work, developed by Ernst Chain, Howard Florey and the often-forgotten Margaret Jennings, saved hundreds of millions of lives. Without it, there is a good chance that my daughter would have died there and then, in the chaos and noise of a paediatric Accident and Emergency room, surrounded by monitors, alarms and sterile bright lights.

I watched the nurse go through her ritual of calculating the dose of the antibiotic. When she was sure, she diluted the sterile vial of antibiotic powder with water and connected it to the intravenous giving set. The pale-yellow fluid passed down the plastic tube and disappeared into my child's arm. The clear plastic wristband denoting her name, date of birth and hospital number was tight against her limp wrist. The antibiotic called Ceftriaxone killed the bacteria that were drowning her, and they destroyed large swathes of Liberty's developing gut microbiome. I felt relief.

I can't tell you how many antibiotic prescriptions I have made in my career; I know it's a very high number. They have been given to almost every patient I have ever operated on. With great antibiotic power comes great antibiotic resistance and I am responsible for playing my part in this tragedy, which was entirely predicted. During his Nobel oration, Fleming stated:

> *The time may come when penicillin can be bought by anyone in the shops. Then there is the danger that the ignorant man may easily underdose himself and by exposing his microbes to non-lethal quantities of the drug make them resistant.*

But he did not predict the global impact of his discovery on our microbiome. No one did.

Antibiotics

I walk past the museum devoted to Fleming every day on my way to work in Paddington, and I have to remind myself that he did not 'discover' antibiotics any more than Columbus 'discovered' the New World. Many antibiotics are produced naturally and have been used throughout human history. In present-day Sudan, for example, traces of antibiotics have been found in Nubian mummies dating back to AD 550. It's thought they were ingested through beer or bread, as the soils used to grow grain contained the bacteria *Streptomyces*. Many foods and natural products also serve as organic antibiotics. The Vedic civilization of northern India, active during the late Bronze Age, revered honey, considering it one of nature's great gifts. It has been used by all major civilizations since then for its wound-healing properties, and is still used today in the NHS. Honey is a complex bioactive substance that serves as, among other things, a natural antibiotic for more than sixty species of bacteria.[2] And Traditional Chinese Medicine (TCM) has used *Artemisia* plants for thousands of years for the treatment of malaria, although the bioactive antimicrobial qinghaosu (artemisinin) was not extracted until the 1970s.[3] Polyphenols from green and black tea, chitosan from the shells of shrimp, bamboo extract and neem oil have all been used as antibiotics, as have minerals and even highly toxic metals like mercury, which was used in the treatment of syphilis.

Until the 1800s we had co-evolved with bactericidal elements from the soil microbiome. The modern antibiotic era began with the discovery of chemotherapy and the birth of what we know today as the pharma industry. Paul Ehrlich, a German physician and polymath mentored by Robert Koch, was at the beating heart of the microbial renaissance. Ehrlich was an extraordinary scientific mind, whose discoveries in haematology, immunology and cancer still influence the medical profession today. He is best known for the discovery of what he called 'chemotherapy', which does not refer to cancer treatment as we know it now. Ehrlich formulated the concept that molecules can specifically bind to cell receptors, acting like a key that can only open the lock it was made for. His goal was to use this theory to create a magic bullet for all disease, and he embarked on a systematic and industrial search for these chemical keys. In 1909, at his 606th attempt, the compound 'Arsphenamine' was discovered.

Arsphenamine was an arsenic derivative modified by Ehrlich's assistant, Sahachiro Hata, into a drug called Salvarsan for the treatment of syphilis. The first agent with a specific therapeutic effect to be mass-produced based on chemical theory, it was arguably the first blockbuster drug. Its second iteration, called neosalvarsan, became the most widely prescribed medicine globally until the discovery of penicillin, which tells you a lot about how STIs have influenced the evolution of the microbiome.

Ehrlich set the scene for the whole pharmacology industry, and he invented modern pharmacology to wage war on bacteria. Without him there would have been no Prontosil.

In the 1940s, while the British were trialling penicillin and putting an end to the neosalvarsan gold rush, an American botanist named Benjamin Duggar discovered a yellow crystalline compound with unique antibacterial properties called aureomycin. He found it in the soil of cemeteries with high concentrations of *Streptomyces aureofaciens*, a fungus-like bacterium of the Actinomycetales order. This was the same mould that found its way into Nubian beer, and it was another monumental rediscovery. *Streptomyces* is the largest genus of Actinobacteria, and it possesses more than 500 species. Today almost all antibiotics in medical use are derived from the Actinobacteria phylum, which means the majority of antibiotics that are mass-produced originate from the soil microbiome. These organisms are prolific producers of specialized metabolites that include the antibiotics streptomycin, tetracycline, chloramphenicol, erythromycin and vancomycin.

Many microbiologists consider Selman Waksman to be the actual father of antibiotics, although I am biased towards Fleming, given that I work in his hospital. Waksman was awarded his Nobel Prize in 1952 for the discovery of streptomycin, the first modern antibiotic derived from Actinobacteria to be effective against tuberculosis. He was also the first to use the word 'antibiotic' as a noun in 1941 to describe any small molecule made by a microbe that antagonizes the growth of other microbes. Not all antibiotics kill other bacteria, and some simply inhibit growth – in which case they are bacteriostatic. With these discoveries the antibiotic industry had at its disposal multiple antibiotics that all worked through different mechanisms. And the Second World War dramatically sped up their manufacture and distribution. The rest is history – except that it isn't.

The resistome

Because bacteria have been in an arms race over resources they have developed antibiotic counter-measures. Many antibiotic agents have a four-atom ring known as a β-lactam (pronounced beta lactam). The first enzymes designed to break this ring and neutralize the antibiotic action, called β-lactamases, appeared about two billion years ago,[4] and the genes encoding resistance to β-lactam, tetracycline and glycopeptide antibiotics have been identified in metagenomic samples of 30,000-year-old permafrost, and in the gut microbiome of a pre-Columbian Andean mummy from Cuzco in Peru (AD 980–1170).[5] The human microbiome therefore served as a reservoir of resistance genes long before antibiotic use became widespread. For as long as humans have been consciously developing antibiotic therapies, our pathogenic and symbiotic microbes have been trying to find ways around them. But the number of antibiotic resistance genes residing in the gut grew exponentially after the modern rediscovery of antibiotics when, amazingly, it was not appreciated that a pathogen could be resistant to multiple antibiotics simultaneously.

Most resistance traits are acquired by bacteria through horizontal gene transfer. Once a common ancestor finds a home in the gut, it's quickly able to share multiple resistance genes with its neighbours by using a large suite of different gene-sharing tricks (for instance, plasmids and integrons), and the entire ecosystem can then access the code. It's like having a cyber-hacker inside us, constantly bypassing the firewall defences of our operating system to spread disinformation among our vulnerable populations of symbionts. The sum of the bacteria living within us that contain resistance genes is known as the 'resistome'.

A major driver for antibiotic resistance comes from the environment and 'Big Ag'. As early as 1951 the first reports of antibiotic resistance were made in animals being used for human consumption, through the experimental feeding of streptomycin in turkeys.[6] In both the East and the West, attempts over the last ninety years to regulate the use of sulpha drugs, quaternary ammonium products and antibiotics in farming largely failed, until the mid-1990s when Sweden campaigned for and won antimicrobial growth promoters (AGP) bans, helped in large part by Britain's mad cow disease (BSE) crisis. Denmark is the world's largest exporter of pork, and in the mid-1990s it also banned Avoparcin, an antibiotic like vancomycin used in food animal production. After this ban, researchers observed that levels of vancomycin-resistant enterococci found in Danish livestock and humans dropped within two years, and Danish pork production even managed to maintain its output. In 2006, the European Union finally banned the use of AGPs in animal food and water, and between 2011 and 2020 sales of AGPs declined by 43 per cent.[7] The veterinary use of antimicrobials also peaked in 2015 in the US and has declined since then. However, antibiotics are still very much used as AGPs all over the world; policies are heterogeneous and are systematically abused because we are addicted to the consumption of meat.

Of course we also give a lot of antibiotics to humans. A recent analysis from the Department of Global Health at the University of Washington analysed worldwide antibiotic prescribing rates. It found that global antibiotic consumption, expressed in defined daily doses (DDD), increased by 65 per cent between 2000 and 2015 to somewhere between 21.1 and 34.8 billion DDDs, and this increase was much higher in low- and middle-income countries.[8] As a result of this orgy of prescribing, the UK now spends about £1 billion each year trying to treat the most common resistant bacteria like *E. coli, Klebsiella pneumoniae, Enterococcus faecium, Pseudomonas aeruginosa* and Methicillin-resistant *Staphylococcus aureus* (MRSA) – all pathogens that no longer respond to the last line of antibiotic defences. And this is just the shortlist; *Acinetobacter baumannii, Neisseria gonorrhoeae, Helicobacter pylori* and *Myobaceterium tuberculosis* also cause havoc. More than half of the patients to whom I prescribe antibiotics during surgery are already resistant to them, and I don't have any others to give them. This means that surgery is going to become a lot more dangerous.

The current global demand for antibiotics must be met by

industry, and pharmaceutical factories all over the world leach antibiotics into our streams, waterways and soils, contaminating every corner of our environment. Patancheru is an industrial zone twenty miles outside Hyderabad in central India, a region with diverse waterways and lakes. Treated waste from antibiotic production and drugs from more than 300 manufacturers are tipped into brown, festering open tanks or streams that supply these lakes. The volume of antibiotic spillage into these wastewater sources is difficult to comprehend; an analysis of waterways from this region found such a high concentration of a broad-spectrum antibiotic called ciprofloxacin that it could have been used to treat 44,000 people.[9] The irony is that the patients treated in clinics and hospitals in this region require larger doses of antibiotics for treatment, sometimes incorrectly prescribed for the chronic inflammatory conditions that are caused by antibiotic contamination, such as eczema and asthma. And the hospitals where they are prescribed return these antibiotic resistance genes back to the rivers. Multiple experiments have identified that untreated hospital wastewater from across the world strongly selects for multi-resistant *E. coli*.[10]

Hyderabad is not an isolated case, and there are examples from all over the developing and developed world of the same phenomenon. As Dr John Snow knew (see page 44), sanitation and waste infrastructure are critical for our health, but in his day sewers didn't contain drugs and antibiotics that also influenced the risk of non-communicable chronic disease. Using metagenomic strategies, we're now able to study antibiotic resistance genes across populations to get a sense of the size of the problem. The international Metagenomics and Metadesign of Subways and Urban Biomes (MetaSUB) demonstrated that the number and type of antibiotic resistance genes found in microbiomes in cities vary widely. Some cities, like Bogota, have fifteen to twenty times more resistance genes than European cities such as Stockholm, which, given its progressive approach to managing this problem, has a low density. As well as reflecting varying antibiotic policy, the MetaSUB findings can be put down to differences in urban geography or to the variation in the baseline microbiome.

By conservative estimates, antibiotic resistance will cause ten million deaths globally by 2050, at a cost of £66 trillion.[11] We are officially in the post-antibiotic era, but the lack of new drugs for common bacterial infections isn't the really big problem.

The data on the gut resistome tell us something profound about the evolution of the human microbiome, which has been transformed in just seventy years. In 1961, Waksman pointed out that naturally occurring antibiotic molecules made by some bacteria may have important signalling roles because they can influence how genes are transcribed in target bacteria.[12] It's only when they reach a critical concentration that they become dangerous to bacteria. We still know relatively little about the ecological role of the molecules that we think about as antibiotics and, from a microbe's perspective, the mass production of antibiotics has created considerable confusion and noise, not only in the bacterial communities that contain our microbial symbionts, but also in those that maintain our planetary health. The problem is not that we've forced pathogenic bacteria down a mutated evolutionary path that has made them more dangerous and even harder to treat; it's the extinction of the symbiotic microbiome that maintains our health and nurtures our planet. When it's gone, it's gone, and it's disappearing at an alarming rate. The resistome therefore serves as a biomarker for the health of our cities, our waterways, our farms and our gut. It is a marker of poverty, conflict and the failure of healthcare systems.

And yet I keep prescribing.

The antimicrobial paradox

Antibiotics saved my daughter's life, and the lives of millions of other children exactly like her. However, this miracle has come at a price. A cumulative exposure to systemic antibiotic therapy has been strongly associated with a greater risk of new-onset autoimmune diseases such as type I diabetes, rheumatoid arthritis and inflammatory bowel disease,[13] and a higher risk of colonic polyps and bowel,[14,15] liver[16] and breast cancer.[17] Antibiotic use in early life is also strongly associated with asthma, allergies, obesity and type II diabetes, and it changes the way our drugs and vaccines work. The true impact of widespread use of antibiotics on the gut–brain, liver and lung axis, the genderome or drug metabolism is incompletely understood, but given the complexity and co-dependence of these systems, antibiotic misuse is likely to be of significant importance to their healthy function. It is therefore important to understand what happens to our microbiomes when we take an antibiotic.

Because everyone's microbiome and resistome are different, this is surprisingly challenging. What's more, antibiotics may be taken alone or in combination, and for differing lengths of time. There are many to choose from and they work in lots of different ways, which means that the influence of antibiotics across symbiotic networks is very variable. We don't just take antibiotics as medicines – they're in our environment and in our food, and working out which specific active antibiotic is driving any change within a niche across a population is hard. Their effect is confounded by many other environmental factors that influence our gut microbiome, such as diet, pets, drugs, breastfeeding and even geography. Finally, many studies that have examined the impact of antibiotics on the microbiome haven't used standardized methodologies to measure their influence. Because of this complexity, remarkably little is really understood about the detail of the impact of antibiotic usage on our symbionts. However, we can say that antibiotic treatment has the potential to disrupt every aspect of the human microbiome: the absolute number of bacteria, the relative proportions of bacteria, the functions of those networks of bacteria, their rate of gene transfer, co-metabolism and general symbiotic functions.

After a short course of oral antibiotics like penicillin for a simple chest infection, the gut will experience a mild but significant reduction in microbial diversity. In most cases, our microbiomes will have enough plasticity to regrow into their pre-treatment structure after a few weeks, but longer courses of multiple antibiotics that are designed to attack a larger number of bacteria (called broad-spectrum antibiotics) can have a devastating influence. In one randomized study of twenty-two humans, those taking a five-day oral cocktail of neomycin, vancomycin and metronidazole caused a 10,000-fold reduction in gut bacterial load, and a change in diversity that in some cases lasted more than six months.[18]

After taking broad-spectrum antibiotics, the relative abundance of the Actinobacteria phylum typically falls, and in particular the genus *Bifidobacterium*. Conversely, the Bacteroidetes phylum generally benefits from antibiotics and its numbers may climb. In some cases, certain species will explode into a massive bloom, which is precisely how the hospital superbug *Clostridium difficile* causes disease – and this is precisely what happened to Raymond (see page xvi). Even when gut diversity doesn't shift, antibiotics are linked to significant changes in our gut metabolism, which can be so significant that it alters our ability to regulate our blood sugar.[19] For people who don't have a resilient microbiome, it might never return to its pre-treatment composition, diversity or function – in these cases, the gut microbiome demonstrates 'antibiotic scarring'. Some individuals are more susceptible to this than others; in one longitudinal study, patients who took a single dose of azithromycin demonstrated similar gut diversity to patients who'd been treated in intensive care.[20]

Antibiotics have the power to change our internal climate because they don't just kill bacteria. They also have an influence on fungi, protozoa and even nematodes and all aspects of the broader microbiome. Antibiotics are well known to influence the balance of phage in our gut, and in turn their functions of lysis and lysogeny on bacteria.[21]

Recent analysis suggests that there are three important factors that influence how taking antibiotics in early life can alter the risk of chronic disease. The first and most important of these is the timing of someone's exposure to antibiotics. The World Health Organization (WHO) reported that in low- and middle-income countries, children under two receive an average antibiotic use of 4.9 courses per child per year.[22] In the developed world, the Centers for Disease Control and Prevention (CDC) reports that children are prescribed almost 7 million antibiotic prescriptions in emergency departments annually,[23] and antibiotics are the most prescribed medications in neonatal intensive care units.[24] But exposure can come even earlier: antibiotic prescription during gestation is especially important because of the importance of the maternal orchestral signalling. Antibiotics can also reach the infant via breast milk. Several studies have found associations between maternal antibiotic use and an increased risk of asthma, wheezing, eczema and atopic dermatitis in babies, although the link with food allergy seems to be less strong.[25] Antibiotic treatment in the first week of life is also independently associated with an increased risk of wheezing and infantile colic, and exposure to antibiotics during the first year of life has been strongly associated with childhood asthma, allergies and changes in infant birth weight, even after adjusting for familial factors.[26]

While antibiotics will have the biggest effect on our microbiome in infancy, taking them as adults can also affect our risk of developing some chronic diseases. UK researchers studied 22,677 patients with rheumatoid arthritis, an autoimmune disease that typically causes painful and debilitating swelling of the joints. They compared this group to 90,013 people without the condition, but of comparable age and sex, and looked back ten years before the onset of the rheumatoid diagnosis.[27] Patients who'd been given antibiotics during this time were 60 per cent more likely to develop the condition than those who hadn't, and this also correlated to the dose or frequency of antibiotic prescriptions. Interestingly, both antifungal and antiviral prescriptions were also associated with an increased risk of getting this disease.

The second major factor is the type of antibiotic. In 2020, researchers from Great Ormond Street published a paper in the *British Medical Journal* that analysed 104,605 children born between 1990 to 2016 to mothers prescribed one macrolide (erythromycin, clarithromycin or azithromycin, for example) or one penicillin monotherapy at any time from the fourth gestational week up until the point of delivery.[28] What they found was disturbing: prescribing macrolide during the first trimester was associated with an increased risk of all types of congenital malformations, including disorders of the cardiovascular, gastrointestinal, genital and urinary systems.

Prescriptions in the first trimester were particularly associated with cardiovascular malformations, and those in the third trimester with urogenital malformations. They didn't, however, find a statistically significant association between the use of antibiotics and neurodevelopmental disorders.

Macrolide use in early life was associated with an increased risk of asthma and a predisposition to antibiotic-associated weight gain. A group of 142 children aged between two and seven were studied in Finland over two points in time.[29] The researchers found that macrolide antibiotics were strongly associated with a long-lasting shift in gut microbiota composition and metabolism – the same association wasn't seen in penicillin use. Those overweight and asthmatic children had distinct microbiota compositions compared to children who hadn't been exposed to the antibiotic, further suggesting a causative link.

Finally, the frequency with which children are given antibiotics also matters. Researchers gave non-obese mice that were genetically susceptible to type I diabetes antibiotics using different dosing strategies. The first mice received antibiotics in separate doses, while the control mice were given continuous low doses of antibiotics. They found that those mice given antibiotics in separate doses experienced greater disruption to their microbial community composition when compared with the controls. This caused more inflammation in their gut because its capacity for lipid metabolism was changed, which in turn meant they were more likely to develop diabetes than the controls.[30] For similar reasons, antibiotic use in infancy predicts obesity risk because the metabolic programming of the gut is profoundly disrupted.[31]

Luckily, antibiotics are not the only tool we possess for killing microbes.

Vaccination

Vaccines are an ancient form of pathogen prevention, and they represent the pinnacle of healthcare innovation. It seems strange to laud the human ability to eradicate highly pathogenic microbes from the planet at a time when we're currently causing a microbial mass extinction. But our ability to selectively prevent a disease before it causes suffering (or even afterwards) is extraordinary. Today, global immunization programmes prevent an estimated two to three million deaths per year[32] and, when we choose to, we can develop a vaccine within a single year for a completely new pathogen. Vaccines have saved us from diseases so successfully that in just a few decades we have forgotten how horrific it was to die slowly from a disease like smallpox, covered in agonizing pustules that drip pus and blood, or how miserable the disability of polio is. Vaccines keep people in work and the economy ticking over, and there is very little robust evidence that they contribute to our risk of chronic disease.

Today, the people of the UK are offered vaccines against diphtheria, tetanus, pertussis, polio, *Haemophilus influenzae* type b (Hib), hepatitis B, rotavirus, meningitis B, pneumococcal disease, measles, mumps, rubella, flu, HPV, shingles, tuberculosis and COVID-19, among others. The WHO recognizes twenty-nine diseases that are preventable by vaccines. We even vaccinate pets to keep them safe, and our livestock healthy for our consumption. We do this because vaccines are incredibly effective. My children are vaccinated, as am I.

Despite overwhelming evidence for their benefits, vaccines remain one of the most divisive medical treatments available to humanity. Vaccination for smallpox became mandatory in 1853 – with a penalty of imprisonment for anyone not complying. By the time Pasteur began publicizing his work on the anthrax vaccine ten years later, the anti-vax movement was already in full effect. Over the last century and a half the debate has raged on, fanned by the flames of conspiracy theories created by legislation that has excluded or punished those who have chosen not to contribute to herd immunity. The biggest motivation for people who decline vaccines is fear, and we can't exchange choice in vaccination status for trust.

All medical therapies have side-effects and risk, and vaccines are no different. Vaccines can, rarely, cause allergic reactions, and the mRNA COVID-19 vaccine caused symptoms such as fever, nausea and diarrhoea, which may have contributed to fatal outcomes in some frail patients. It was also associated with a minimal risk of cardiac inflammation in the young.

To date, all evidence suggests that vaccines don't cause chronic disease and there's very little evidence (if any) that vaccines have significantly changed the ecology of the human microbiome. Unlike antibiotics, vaccines work in a targeted way that leads to fewer casualties among the symbiotic population of the human microbiome. If they do alter the human microbiome, it is certainly not at the scale of antibiotics. Unlike antibiotics, vaccines are not ubiquitous. On the contrary: there haven't been enough vaccines to go round, and inequality of access is the bigger problem.

There is also very good evidence that vaccination reduces antibiotic usage and, as a result, prevents antibiotic resistance – and that it contributes to our herd immunity for many diseases. For example, pneumococcal conjugate vaccines and oral rotavirus vaccines prevent an estimated 23.8 million and 13.6 million episodes of antibiotic-treated illness respectively, among children under five in low-income and middle-income countries each year.[33] During the COVID-19 pandemic the relative number of antibiotic prescriptions climbed by 6.7 per cent in the NHS before the introduction of the vaccine. From a human ecological perspective, vaccines directly and indirectly limit the destruction of our microbiomes; their manufacture is also kinder to the environmental microbiome.

Antibiotics also have a direct impact on the effectiveness of vaccines because they change the microbiome. For example, infant mice given antibiotics have an impaired antibody response to at least five different types of licensed vaccines that are currently administered to infants across the globe.[34] This vaccine resistance can also be reversed by administering a faecal microbiota transplant to antibiotic-treated mice, from age-matched, untreated mice. Antibiotics in humans also reduce vaccine efficacy, although it is not quite so straightforward; researchers from Stanford gave large doses of antibiotics to twenty-two adults before and after their seasonal flu vaccines; only those adults with existing low levels of antibodies for the H1N1 flu virus had a significantly impaired antibody response to the vaccine when given antibiotics. This probably reflects the fact that immune responses in adults to vaccines are mainly determined by immune history and they are resilient to transient changes in the microbiome, although this may not be the case in infants.

The microbes in our gut have many other mechanisms through which they could interfere with vaccine efficacy, as they produce proteins, toxins and antigens that directly interfere with the immune machinery. If the microbiome is important in defining the efficacy of vaccines, this is vital information, because its functions are modifiable by a large range of environmental stimuli, and none more so than diet. A high-fat diet will shift the entire ecology of the microbiome and its metabolic functions, such as the production of bile acids. The Stanford study also noted the major impact that antibiotic use had on circulating bile acids, associated with increased inflammation. This makes Krispy Kreme's offer of free doughnuts for those taking up COVID-19 vaccines in the US during the pandemic seem counterproductive.

All of this may help explain the wide global variation in vaccine efficacy – which maps on to global variations in the gut microbiome. Protection against tuberculosis by the Bacillus Calmette–Guérin (BCG) vaccine, for example, varies from 60 per cent to 80 per cent,[35] with a higher response rate in Europe than in Africa. And vaccines against poliomyelitis, rotavirus, malaria and yellow fever provide less protection in Africa and Asia compared with Europe or the US. Microbiome studies in this area of research do show conflicting results, depending on the vaccine and the method used to study the microbiome. But there are consistent observations for those being vaccinated in urban environments and those in low- and middle-income countries, where microbiome variation is at its greatest and vaccine effectiveness is at its lowest.

This creates something of a paradox. The typical rural African gut microbiome is more diverse and is populated with species of bacteria that promote fibre metabolism and prevent inflammation in the gut. So why wouldn't this far healthier microbiome respond to the vaccine? This may be the point. It is possible that when the gut microbiome is especially diverse, the body's immune defence is simply too good to be susceptible to synthetic vaccines trying to subvert it. This does not mean that intestinal biodiversity should be sacrificed for vaccination with antibiotics, but that we need a deeper understanding of this relationship, and we need vaccine programmes that are optimized for the specific populations they have to work in. All of this emerging evidence has significant implications, not only for the deployment of the COVID-19 vaccine, but for all vaccination programmes around the globe where access and biological response are likely to vary.

Microbial peace

As I sat on the seventh floor of St Mary's Hospital watching my pale, breathless daughter recover from her pneumonia, I didn't care about any of this and I was merely grateful for the antibiotics. And I'm all too aware that they're a precious and rapidly diminishing resource. It's even possible that my daughter might not be able to give them to her children.

No matter the antibiotic choices that we make for our offspring, their microbiomes will inevitably be shaped by them. Seventy-five years of blissful manufacturing, compulsive prescribing and wilful misuse in farming have caused a dramatic dysfunction in global environmental microbial ecosystems. This affects all life forms on Earth, and it's had unintended and deleterious consequences for our health, some of which may be impossible to reverse. But we are not concerned with this: instead, we're fixated on finding ever-stronger drugs to kill more bacteria so that we can continue this destructive cycle. This policy entirely misses the point of the microbiome.

'Abiosis' describes a life without bacteria. This is simply impossible, and it cannot be our objective. A new strategy is needed that doesn't see microbes as the enemy. We have to move towards a form of holistic healthcare that prioritizes biodiversity within all living creatures, particularly those that we rely on for our coexistence. A core component of our health policy should acknowledge the benefit of symbiotic microbial life forms to health. We need to move away from blunt instruments of destruction that cause massive collateral damage to our internal ecology, towards a world where we can selectively target pathogens while promoting important symbionts. Vaccines will be one vital tool in achieving this goal, but for now we have to rely on expensive drugs.

12. *The Drugs Don't Work*

After prescribing a new medicine for a patient, the first question I will ask on seeing them again is, 'Do you feel better?'

It's a strange question, when you stop to think about it. The medicine was expensive, it was synthesized, it was trialled, it was 'clinically proven'. Many times the answer is a resounding 'yes'. Often the answer is 'no', 'I don't know' or even 'I feel worse'. You wouldn't buy a plane ticket if the aeroplane only landed some of the time.

The number of drugs prescribed globally every year is mind-blowing. In 2020, the world's population received 4.5 trillion doses of medicine, up 24 per cent from 2015 (it's a lot, even when we factor in COVID), and over half of the world's population will consume more than one dose of medicine per person per day, which is an increase from one-third of the world in 2005.[1] This has been driven by massive growth in the consumption of medicine in India, China, Brazil and Indonesia. As a result, global spending on medicine will reach US $1.6 trillion by 2025.[2]

It's bizarre that we spend all this money on drugs that don't cure our chronic disease. In 2019, the most prescribed class of drugs in the NHS were 'statins' taken for lowering cholesterol (77,997,663 prescriptions). Globally in 2021 the most profitable drug was a medicine called Humira, an antibody used in the treatment of auto-immune diseases like rheumatoid arthritis and inflammatory bowel disease; it made the pharmaceutical giant AbbVie approximately US $20 billion. Both of these medicines are designed simply to keep these diseases in a very profitable remission. There's no money in truly alleviating human suffering – as a result, we have an epidemic of diseases of progress and no effective cures.

Pharma is a high-risk, high-return business. At a conservative estimate, it costs on average US $1.3 billion and twelve years to take a drug from bench-side to clinic.³ And it has a high failure rate; out of every ten new drug candidates, nine will fail to reach the market. The biggest problem of all, though, is that the pipeline of new drugs is running dry. Pharma companies have now moved from a business model of carpet-bombing markets with single drug candidates to a more nuanced strategy that accounts for individual variations in biology, based on genomic analysis. The objective of 'personalized medicine' is getting the right drug into the right person at the right time for the right reason. It is as much about improving the discovery hit rate, lowering costs and maximizing profit as it is about drug safety and improved outcomes.

Microbiome science tells us that the entire basis for drug discovery is flawed, because it doesn't account for the influence of the microbiome. Pharma didn't realize that you cannot have personalized drug treatments without considering the most important source of human biological variation – namely, the microbiome. Then again, pharma has a history of not learning its lessons, and there are lots of examples where the importance of the microbiome has been missed.

The antibiotic drug Prontosil, manufactured by the German company Bayer prior to the Second World War, failed in its original design because it was converted within the body to its active agent, sulphonamide. As it happens, this conversion was driven by gut bacteria – the very thing the drug was designed to kill.

Pain

As soon as the Girl and I had children, a drug called paracetamol (or Tylenol, as it is also known in North America) became one of the most precious elixirs in our parental armoury. If you time it right, the thick, sweet purple liquid of its paediatric formulation (known in the UK by its commercial name of Calpol) can be syringed into the mouth of a pink-faced, febrile child between angry sobs, and the impact is transformational. In adults, the same drug cures the headaches of exhausted parents and, without it, boardrooms all over the world would fail and hangovers would be fatal. Today, paracetamol is one of the most widely studied and prescribed drugs anywhere in the world. But despite its ubiquity, we don't know definitively how it works.

The drug readily crosses the blood–brain barrier, and its ability to lower a child's temperature is most probably the consequence of its influence on multiple pathways (including the endocannabinoid, serotonergic and nitric-oxide pathways); it also eases our pain by blocking the production of prostaglandins that cause inflammation, through the inhibition of an enzyme called cyclooxygenase (COX). Because you can buy this drug in any general store across the world, you would be forgiven for thinking that it is a safe drug, but you would be quite wrong. It causes high blood pressure and gastric ulceration, inhibits blood from coagulating and increases our risk of heart attack. It also comes with a low risk of precipitating so-called pseudo-allergic reactions, including asthma, and inhibits prostaglandin production during pregnancy that can influence foetal health. Most importantly, the overconsumption of paracetamol can be fatal. It's responsible for 56,000 emergency department visits, 2,600 hospitalizations and 500 deaths per year in the US, and almost 50 per cent of these are unintentional overdoses. If this drug were discovered today, it would not be licensed for use.

So why do humans respond so differently to this over-the-counter painkiller? The liver does the bulk of the work when it comes to metabolizing medicines. The 1950s saw a breakthrough in pharmacology with the discovery of a superfamily of enzymes in the liver responsible for the metabolism of many xenometabolites and drugs like paracetamol, called cytochrome P450 (CYP450). Because this family of enzymes is so large and mutations in the genes that code for these enzymes are so common, it was thought to explain much of the human variation in drug metabolism. If, say, you have one mutation in a particular type of enzyme, it might make you more vulnerable to the toxicity of a drug like paracetamol. But these enzymes – and others like them that break down the secondary metabolite of paracetamol in the liver – work variably in those who are malnourished, sick or obese: conditions associated with an altered gut microbiome function.

We take many of our medicines by mouth, and so it follows that they'll come into direct contact with microbes in the gut as they are absorbed, and these microbes possess a large toolkit for altering the chemistry of these pharmaceuticals. It's surprising, then, that until recently the microbiome was not thought to be an important player in the field of pharmacokinetics, which is the study of how a drug is absorbed, distributed, metabolized and excreted. And we still don't fully appreciate the fact that drugs are metabolized by the microbiota in the gut. Microbes can deactivate or reactivate medicines before or after they have passed through the liver, and because they have the keys to our immune system, they can even influence the effectiveness of medicines given intravenously (like chemotherapy).

A study of how humans metabolize paracetamol, conducted at Imperial College London, discovered that people with high levels of a microbial co-metabolite (a metabolite produced by a combination of human and bacterial metabolism) called *p*-cresol, which is found in the urine, are less able to metabolize paracetamol and are more vulnerable to its side-effects. That is because this co-metabolite is generated by the *Clostridia* species of bacteria in the gut, which require lots of sulphur to make it. The enzyme in the liver that detoxifies paracetamol is also dependent on sulphur, which means that individuals who have too many *Clostridia* species cannot safely detoxify paracetamol. Those people who eat a high-fat Western diet are far more likely to have *Clostridia* in their guts. While the microbiome influences the toxicity of the most frequently taken medicine on Earth, we have no measure of its impact in routine use. The implications of this observation for how we take drugs cannot be overstated.

Paracetamol is not the only painkiller with which the microbiome interacts. Between 1999 and 2019 nearly 500,000 people died from an overdose involving any opioid in the US, and more than 1.6 million people now suffer from opiate dependency.[4] Recent revelations that the Sackler family – the billionaire owners of Purdue

Pharma – helped trigger the opioid epidemic with dishonest marketing of their opioid OxyContin have caused outrage,[5] although their egregious advertising of antibiotics for greedy pharmaceutical companies (and the failure of regulators to stop them) may well have done even more damage to society.

The social and economic collapse within local communities blighted by addiction perturbs the microbiome. Studies of the gut microbiome in people who suffer from substance abuse have demonstrated a reduced species diversity, and markedly different abundances in bacterial taxa when compared to healthy controls.[6] Opiates directly change the ecology of the gut microbiome even when dietary and lifestyle confounders are accounted for, and animal studies suggest that changes in diversity can occur in just one day. This is important because the microbiome is, in part, responsible for the bioavailability of opiates, which may contribute to the tolerance of its painkilling effect.[7] Evidence for this comes from studies of the psychobiotic strains of *Bifidobacteria* and *Lactobacillaeae*, which have been shown to correct the resistance to pain control that regular users of these drugs experience.[8]

However, this does not necessarily explain how the microbiome contributes to morphine-induced rewarding and reinforcing behaviours that are seen in addiction. Some clues that explain this observation are coming from germ-free mice experiments and mice given antibiotics. Both of these groups of animals are less intolerant to opioids, and the microbiome has multiple pathways through which it can modify addiction susceptibility. For example, oxycodone is a strong synthetic opioid used to treat severe pain, and microbiome depletion with antibiotics influences the activation of neuronal ensembles in the regions of the brain that drive addiction.[9] Faecal transplantation of these antibiotic-treated mice with a control microbiota restores these behaviours,[10] suggesting that this may be a valid therapy for helping addicts overcome this destructive condition.

There are examples of the microbiome being leveraged to treat other substance addictions in this way. For example, a randomized placebo-controlled trial of FMT in twenty patients with alcohol dependence syndrome showed that it reduced cravings for alcohol in 90 per cent of those people given the FMT, compared to 30 per cent in the placebo group. The FMT cohort also had improved cognition and psychosocial quality of life.[11]

This work is incredibly important to me, as a surgeon, because I need good pain control for my patients, and the microbiome may yet help me improve the safety and effectiveness of the drugs we use to achieve this. Historical and current epidemics of drug addiction are a catastrophe of our own making, and those whose lives have been destroyed by addiction deserve sympathy and help. The microbiome offers new strategies for addiction where very few effective treatments exist.

Chemotherapy

Chemotherapy for the treatment of cancer is effectively a poison. It targets the process of cell replication to stop the uncontrolled cell growth that characterizes tumours – the reason cancer kills. Chemotherapies come in many different forms, even including some types of antibiotics, and they typically target some part of the DNA replication process. They can be given at different times relative to surgical treatment and in combination, so unpicking their relationship with the microbiome isn't straightforward.

A major problem with cytotoxic chemotherapy is that it works indiscriminately, meaning that it also influences healthy cells, leading to a series of dreaded side-effects such as intestinal mucositis (inflammation of the bowel), myelosuppression (inhibition of the white blood cells in bone marrow, or neutropenia), alopecia (hair loss) and bacteria. About 40 per cent of patients receiving standard-dose chemotherapy and 100 per cent of patients receiving high-dose chemotherapy exhibit some symptoms of nausea, vomiting, abdominal pain, diarrhoea and malnutrition associated with intestinal mucositis. Almost one in ten cancer patients will require hospitalization for an infection during chemotherapy, and in 85 per cent of these patients, sepsis is caused by organisms that originate in the gut.

The microbiome plays an important part in determining how and when these symptoms develop. First, chemotherapy changes the structure of the human gut microbiome, and it's characterized by reductions in diversity and richness. Studies in humans suggest that different chemotherapies do this in different ways. Irinotecan, a drug commonly used in the treatment of bowel and lung cancer, causes increases of *Clostridium* cluster XI and Enterobacteriaceae populations in the gut, both of which are potentially pathogenic. This drug causes severe diarrhoea in up to 40 per cent of patients, and they are then either given a lower dose or, in many cases, prematurely taken off the drug altogether. However, this does not occur because of too many of the wrong bugs. Instead it's caused by a bacterial enzyme called β-glucuronidase, which reactivates a metabolite of the drug in the gut, causing gut inflammation. The good news is that these enzymes can be blocked to prevent the symptoms.

Chemotherapy-induced peripheral neuropathy (CIPN) describes painful nerve damage and numbness in the fingers and toes of patients given certain chemotherapy drugs that are used to treat ovarian and breast cancers. There are no cures for this very common side-effect and the precise cause is not known. But it's notable that chemotherapy-induced neuropathy is reduced in germ-free mice and in mice given antibiotics before their chemotherapy.[12] When an FMT is performed to restore the microbiome, the neuropathy is restored, suggesting that this side-effect is mediated by bacteria that trigger an immune response through a particular receptor on microglial cells of the nervous system called TLR4. Although this doesn't explain the entire mechanism, it does mean that we at least have a target for a painful side-effect of chemotherapy.

For many patients undergoing cancer treatment, the most dreaded symptom is hair loss. The scalp has its own microbiome because it exists in a humid environment protected from UV light and with a pH favourable to the growth of bacteria. In recent years, research has focused on the role of the microbial community inhabiting the scalp on hair-growth disorders. Very little is known about chemotherapy–microbiome interactions. But there's mounting evidence that the microbiome is likely to be important, driven by small numbers of case reports that have described men with alopecia universalis who experienced sustained follicular growth in response to faecal transplantation.[13] The mechanisms aren't defined, and while it's tempting to speculate that this is another example of microbiome–immune system interactions, for now the jury is out.

Perhaps more importantly, the microbiome also determines how effective chemotherapy treatments are at curing the disease they're intended for. Bacteria have been found within cancer cells for many years, and oral pathobionts such as *Fusobacterium nucleatum* are promiscuous passengers that turn up in a large number of different cancers; they're even present in metastatic deposits in the liver that really should be sterile. *Fusobacterium nucleatum* is responsible for driving resistance to chemotherapy in the treatment of bowel cancer.[14] Just like antibiotic resistance, it's theoretically possible that we can transmit a similar resistance to cancer therapy, which is a frightening prospect.

The great hope of cancer therapy is that we can begin to engineer the body's immune system to do away with these drugs completely, and it is this that is really driving the interest in the microbiome for cancer therapy.

Immunotherapy

In 1891, a thirty-five-year-old Italian immigrant to New York named Signor Zola developed an aggressive life-threatening sarcoma the size of a chicken's egg on his right tonsil. William Coley, a handsome twenty-eight-year-old surgeon in New York City, injected him with a streptococcal bacterium every three to four days. This was grim work; Zola was in pain and unable to swallow, pouring mucus out of his mouth and nose. Coley's aim was to trigger a severe erysipelas, or bacterial infection of the skin. He had come to the conclusion that this was a good idea, after performing an exhaustive analysis of the records in his hospital after failing to save the life of seventeen-year-old Bessie Dashiell. He had noticed that patients who developed infections of the skin had a survival advantage in the treatment of sarcoma. But despite months of injections into the increasingly desperate Signor Zola, the tumour kept growing back. And then Coley had a brainwave: maybe the strain he was using was simply not aggressive enough. He got in touch with Robert Koch and asked him to send some more aggressive strains of bacteria from Germany, which the celebrated microbiologist duly did. Coley injected the new bacteria into his patient's neck, and within an hour Zola developed chills, pain, nausea, vomiting and a high fever. The attack lasted ten days and the tumour was gone. Zola would live for another eight years before succumbing to his cancer.

Coley went on to inject more than a thousand patients like this, and he had a 57 per cent response rate in the treatment of sarcoma, which was not bad for a cancer that at the time was almost always fatal. The American Medical Association was sceptical, however, and published an editorial in 1894 discrediting the therapy. It was very, very wrong.

The second most profitable drug sold in 2020 was a medicine called pembrolizumab, which is used in the treatment of a skin cancer called melanoma, and in many other cancer types. This blockbuster drug is revolutionizing the treatment of diseases that until very recently had an awful prognosis. It represents a new type of therapy referred to as 'immunotherapy', which aims to reprogramme the immune system so that it targets cancer. These treatments have been designed and tested without any thought at all for the microbiota. Pharma didn't need to think about the microbiome for one important reason. A moderate gain in survival for patients with metastatic cancer is enough for pembrolizumab to make US $14 billion per year.

Immunotherapy works by changing how T cells detect and kill cancer. For example, PD-1 is a checkpoint protein that exists on T cells; it's a sort of 'off switch' that keeps T cells from attacking other cells in the body. PD-L1 is a protein key that keeps this switch in the off position, so when PD-1 binds to PD-L1, it tells the T cell to leave the other cells alone. Some cancer cells have large amounts of PD-L1, which means they can hide from immune attack. Drugs like pembrolizumab are called 'PD-1 checkpoint inhibitors' and they flick the T-cell kill switch, allowing T cells to detect and kill cancer cells. The problem is that these drugs don't work on all cancers, and they're only effective in tumours with a high number of gene mutations, which act as tumour-specific antigens that the T cells can detect.

Cancer treatment of all types is also defined by an interaction between the genome of the cancer, the microbiome and the immune system through a dynamic exchange of immune-signalling molecules. Much of this activity doesn't happen in the cancer itself, but in the tissue around it, known as the 'tumour microenvironment'. This is rich in blood vessels, immune cells and bacteria. Until very recently the importance of bacteria to the microenvironment, and hence cancer immunotherapy, was not appreciated.

Animal research remains a cornerstone of the development process of drugs such as this, because even in the twenty-first century you can't put a drug into a human until you know a little bit about how it performs in a real biological system. So there is a big market for genetically engineered animals. Genetic mutations can be ordered from a macabre shopping list to simulate just about any human condition. The whole point of this is that any two mutated animals are genetically identical, so that when you want to test a drug candidate, such as a new immunotherapy, those drugs will behave in exactly the same way in both animals, if purchased from any lab in the world. During drug experiments, mice or rats are typically kept in cages in small numbers. But these animals are coprophagic (they eat poo) and of course they come into contact with each other. And as you might imagine, they share bacteria. Until very recently no one really thought this had a big impact on drug function. At least they thought there would not be a big variation between cages, or even between animal houses that produce them. However, a study on a 'PD-L1 checkpoint inhibitor' changed all that.

Researchers in Chicago acquired mice from two different mouse facilities, Jackson Laboratory (JAX) and Taconic Farms (TAC), to test the drug. These are genetically identical animals and should have had an identical response to the drug when a melanoma was induced. But instead tumours grew much more aggressively in the TAC mice.[15] This was because the bacteria in the guts of these mice regulated the T cells, which in turn influenced how effective the drug was. Amazingly, when faeces from the JAX mice was transplanted into the TAC mice, this effect was reversed.

The scientists managed to identify a specific *Bifidobacterium* species, which when transplanted alone improved tumour control to the same degree as the PD-L1 checkpoint inhibitor. Yes, a single strain of bacteria was as effective as the billion-dollar drug in limiting the size of the tumour. Subsequent studies in humans have demonstrated that responders and non-responders to PD-1 checkpoint inhibitors have different microbiomes, with (in one Chinese study) greater abundances of *Parabacteroides distasonis* and *Bacteroides vulgatus*. When the microbiomes of responders are transferred to non-responders in mice, hey presto, the drug works.

The microbiome has a large number of different tools at its disposal to influence the tumour microenvironment. It's now possible that we can engineer the microbiome to 'turn on' those cancers that are currently hidden from the immune system. This also means that anything in the exposome – such as diet, or other medicines that interact with bacteria and the immune system – has the potential to influence how these drugs work. This is the reason why diet and the microbiome now represent the new frontier in cancer therapy.

The microbial dosette box

The humble mould has not only given us the gift of antibiotics, it has also stopped us from having heart attacks.

In 1966, the Japanese biochemist Akira Endo moved from his home in Japan to New York to study lipid metabolism. When he arrived in New York, obesity rates were starting to soar, and by 1971 he speculated that microbes would produce antibiotics that inhibited a key enzyme in the production of cholesterol. This enzyme is called HMG-CoA reductase, and it functions as a defence mechanism against other microbes. The search took two years and involved the analysis of 6,000 strains of microbes. But it was worth it, because from the blue-green mould of *Penicillium citrinum* Pen-51, Endo's team isolated a new molecule called Compactin. In 1977, this was successfully used for the first time in the treatment of an eighteen-year-old patient with an inherited form of hypercholesterolemia. The golden age of 'statins' or lipid-lowering drugs had begun, and the *Penicillium* mould had delivered another wonder-drug to the world that was worth billions of dollars.

Billions of people across the world now rely on the drug to help control their bad cholesterol (LDL), yet individual responses are highly variable. It's not clear precisely how the two-way interaction between the microbiome and statins works, but it is happening: cholesterol biosynthesis is also important for microbial health and communication and, as Endo described, microbiota use it as a weapon. Statins change gut microbiome diversity, driving up *Bacteroides* and lowering the proportion of *Faecalibacterium* and other microbiota with low cell densities. These drugs also alter bile-acid metabolism, which is intimately involved with bacteria in the gut.

In practice, we rarely leave the doctor's surgery with a prescription for just one drug, particularly if we have high cholesterol. As we get older, the chronic diseases mount up. By the age of eighty-five, around one in five people have at least two long-term conditions, and about half of seventy-five-year-olds take five or more medicines. This is known as polypharmacy. As well as the economic and quality-of-life implications this entails, a dosette box is potentially dangerous, as many medicines interact with one another, and the body is less able to manage them.

The MetaCardis study is a cross-sectional analysis of 2,173 European residents with metabolic syndrome. Through it, we've discovered that the impact of polypharmacy on the gut microbiome is so great that it can exceed changes caused by the underlying disease. But this isn't always bad; the study also demonstrated that there are additive effects of multiple drugs, which can sometimes shift the microbiome towards a more protective structure and function. Most drug interactions commonly known by doctors make no account for the gut microbiome or the ageing gut, so it's exceedingly difficult to make accurate predictions about how a drug will actually behave in a person before it's given to them.

After cholesterol-lowering drugs, the next most likely to be dumped on to the repeat-prescription list is a class of medicine called 'proton-pump inhibitors'; 24,674,821 of these medicines were prescribed in the NHS in 2017, and they work by blocking the secretion of acid (or H^+ ions, called protons) in the stomach. They're taken to prevent gastric reflux and ulcers (I couldn't have got through medical school without them). They're also commonly given with antibiotics to eradicate the bacterium *Helicobacter pylori*. If you change the acidity in the gastric stomach, you change the type of bacteria that live all along the gut, which explains why regular users (like Raymond, see page xvi) have a slightly increased risk of *Clostridium difficile* infection, and gastrointestinal side-effects are commonly reported. But these drugs are also associated with an increased risk of gastric cancer,[16] inflammatory bowel disease,[17] chronic kidney disease,[18] stroke[19] and even dementia.[20] *H. pylori* also changes the effectiveness of L-dopa, a drug commonly used to treat Parkinson's, which is known to be co-metabolized by gut bacteria.

Many of the more routine drugs we take to manage our chronic disease have an indirect influence on our microbiome, and we're slowly beginning to unravel the complexity of these interactions. A group from Heidelberg University screened more than 1,000 marketed drugs against forty representative gut bacterial strains and found that almost a quarter of the drugs that were designed purely for human targets inhibited the growth of at least one strain under laboratory conditions. And there is massive variation within single strains of the same species. *Bacteroides thetaiotaomicron*, for example, is able to metabolize forty-six drugs, including the anti-hypertensive medicine diltiazem. Because there's such significant variation between the functions of individual strains of this particular bacterium from person to person, there is also huge variation in how individuals metabolize drugs, and we have no measure of it.

It's not just the type of medication intake that impacts on the gut microbiome, but also the dosage, drug combinations and previous exposure to antibiotics that significantly influence our ability to safely break down medicines. These factors are especially variable in the elderly, who remain vulnerable to drug interactions and toxicity. There is, however, another important variant in drug effectiveness.

The microcebo effect

In 1763, Alex. Sutherland, MD, published a book entitled *Attempts to Revive Antient Medical Doctrines*. In the first of six volumes, 'Of Waters in General', he dismissed as 'placebos' the fad medical therapies adopted by physicians. Today, the world is full of placebo doctors, pushing everything from water fasts, full liquid diets and water drunk out of copper vessels. Ten years after Sutherland published his book, a Scottish physician called William Cullen gave an external application of mustard powder to a patient, despite the fact that he knew it was of no clinical benefit. The patient improved and the 'placebo effect' became an instant source of fascination.

Today, the most robust trials take account of the placebo effect by using meticulous study design, and by adding a positive or negative control to the intervention. The aim is to distinguish between the placebo effect and the clinical impact of a new treatment, and therefore to understand its true benefit. It is still a surprise when the placebo outperforms the genuine intervention. If a placebo causes harm, it is then referred to as a 'nocebo', and studies are roughly as likely to produce nocebo effects as they are placebo effects.

The underlying causes of the placebo effect are much broader than bias alone – they incorporate cultural context and human behaviour. The science of the placebo effect sits right at the intersection of immunology and metabolism, neurology, endocrinology, psychology, social learning and human connection (it's often used to explain how therapies such as acupuncture and mindfulness work). The microbiome, too, works across these domains. The fact that the microbiome influences the function of our major organs, and that it intimately connects us to each other and to our environment, is important. So is the fact that it's highly variable between people, and that it is responsive to sugar pills and sham interventions through the gut–brain axis. The administration of inert substances may be incredibly subtle, but they do activate the endogenous opioid system and the endocannabinoid system. Both placebos and nocebos modulate the synthesis of prostaglandins, which are the focus of anti-inflammatory drugs, and neuroimaging studies have shown that placebo analgesia and opioid analgesia share a common neural mechanism.

The microbiome doesn't explain the placebo effect in its entirety, but it must play an important part in determining why some interventions have unexpected outcomes. It's also likely to explain why holistic approaches to healthcare that preach nebulous terms like 'balance' do, in fact, work. And as much as evidence-based hardliners don't want to believe it, inclusive approaches to healthcare that promote warmth, connection and understanding sometimes work better than pharmacy, even though we don't understand why. Then again, we don't understand precisely how paracetamol works. No one deserves pious fraud, and we must de-stigmatize medical treatments that many people in the developing world rely on for their primary source of healthcare. We can only achieve this if we fully understand the mechanisms through which all healthcare interventions work, and for now there is as much to be learned from the placebo and nocebo effect as there is from the intervention. It's time for the 'microcebo' effect to be understood.

13. The Microbiome Café

All of life can be seen in a hospital canteen. There are last suppers and first dinners, leisurely lunches and snatched breakfasts. Exhausted nurses with thousand-yard stares pick the leftovers from last night's dinner out of their plastic containers, while junior doctors dine on Red Bull and crisps (I most certainly did). Patients teeter on their crutches while reaching for processed-meat sandwiches and chocolate, and the canteen staff daydream about the end of their shift and a life far away. Everyone drinks coffee.

On the wards, bedside tables are littered with fizzy drinks, cakes, sweets and, occasionally, fruit that sits untouched, slowly decomposing. Everything is wrapped in plastic. The understaffed and underpaid hospital catering teams carefully plan how to spend the £8.77 per day (it can be as little as £3.80, depending on the hospital) they have for each patient and offer up menus of cold soups, burnt toast and bland, colourless slop. Doctors make orders for 'nil by mouth' while dieticians meticulously plan their patients' feeding strategy, calculating calorific requirements, macro- and micronutrient demands, vitamins, consistency and route. No one in the hospital feeds the microbiome.

Dinner is served

Much of the hard work of digesting food in the gastric stomach and the small bowel is done by human rather than bacterial enzymes. There we absorb nutrients from ingested foods and release hormones that regulate our feelings of satiety, our blood-sugar levels and our gut physiology. This hormone-dependent metabolic network has historically dominated nutritional research, and the microbiome's contribution to nutrition has been viewed as minor.

Until now, food has gone into something of a black hole, full of hungry dark matter, and we make inferences on the requirements of our microbiota based on a handful of metabolic pathways that have dominated twentieth-century food science. For example, bile acid (fat) or short-chain fatty-acid metabolism (fibre).

The restaurant hypothesis of the gut suggests that each microscopic niche of the long tube that is the gut represents a distinct fine-dining opportunity for our microbiota;[1] at each one they can find their optimum nutrients and avoid competition from other greedy diners. In practice, this doesn't quite work as an analogy, as the same species of microbe can dine in two different restaurants at the same time. Our gut microbiome also works as a mutualistic network that shares the leftovers of its meal along a giant chemical superhighway. As a result, many diet–microbiome interactions can appear conflicting or contradictory.

At a species level, microbiomes remain relatively stable in adults with daily variations in dietary intake, but at a functional level, the gut microbiome is constantly fluctuating. This is because the microbiome engine adapts its core metabolic functions across species to meet our nutritional requirements. Most of the time we maintain enough functional reserve in this system so that it can adapt to our occasional indiscretions at our favourite pastry shop. However, we can readily exceed them if we adopt a radical change in our diet, like becoming vegetarian after a lifetime as a meat-eater, or if we pursue more minor changes over a long enough time period.

I'm inevitably asked what it is that we should be eating to optimize our gut microbiome. This is a hard question to answer because the interaction of your microbiome and your diet is as much about what you consume as how much you consume, how you cook it, what else you consume it with, where your food is farmed and how it is reared. Each plate of food will mix food types of a different molecular structure and substrate, and two tins of beans bought from two different manufacturers may be considered differently by our gut microbiome if there is even a tiny difference in their ingredients – for example, in the type of additive. From an evolutionary perspective, we use diet to regulate the microbiome by limiting critical substrates they need for growth. Rather than a Michelin-starred restaurant, the gut is a bustling and noisy café with a million counters and smells, and trillions of hungry diners fighting for their next meal because their survival depends upon it. This rowdy, rambunctious rabble of microbes starts banging the table as soon as you start to think about those beans, and it influences your experience of food from the first mouthful.

A matter of taste

Our gut bacteria don't only provide us with nutritional benefit – they also influence how our food and drink taste. The human tongue detects five types of taste: sweet, bitter, salty, sour and umami, although some argue there is a sixth taste for fat. The tongue possesses a permanent niche of bacteria made up of *Staphylococci* and *Streptococci*, as well as other bacteria that can create sticky biofilms of mucus-like material to form a physical barrier that limits the access of taste molecules. Bacteria in the mouth also metabolize our food and influence the concentration of flavours, thereby influencing taste sensitivity. Taste receptors that detect sweet flavours are responsive to microbial chemical signals (known as quorum signals), lipopolysaccharides and toxins. This might be a design feature to protect us from pathogens, because it activates the secretion of saliva, which contains bacterial antibiotics and also triggers hormone secretion in the gut. In severe cases, these same molecules can also induce programmed cell death in tastebuds, leaving us with no taste at all with which to enjoy food or drink.

Therefore we experience unusual tastes and an altered appetite when we're ill, or if we take antibiotics or medicines. This also explains why some people with COVID-19 experienced a loss of taste and smell. The SARS-CoV-2 spike protein binds to taste receptors and hacks into our tastebuds, where the virus replicates, destroying the nerves that carry taste signals to the brain. However, the severity of COVID-19 infection was also associated with overabundances of bacteria such as *Coprobacillus*, *Clostridium ramosum* and *C. hathewayi* that are commonly found in the gut and mouth.[2] Those people who got COVID-19 and lost their taste may have been disadvantaged by their oral microbiome because these bacteria modify the inflammatory response to the SARS-CoV-2 virus.

Our gums, teeth and dental plaques harbour bacteria that have been found all over the body, such as in our furred coronary arteries and in cancers. For example, *Fusobacterium nucleatum* is found in all cancers in the gastrointestinal tract, and even in breast cancer. Debate rages as to whether the bacteria cause these diseases or whether they serve as a biomarker for their presence. We know for certain, though, that oral health is as much about improving our diet as it is about brushing our teeth.

The good news is that we now have access to lots of delicious foods to eat, from across the globe, at any time we want them. This is changing our microbiome, and not always for the better.

The global microbiome kitchen

After the Second World War the UK faced massive labour shortages as it sought to rebuild its shattered economy. In 1948, the government gave Commonwealth citizens free entry to Britain. The arrival of HMT *Empire Windrush* in June of the same year was the beginning of a mass migration event, with hundreds of thousands of people arriving from the West Indies, Pakistan and India to make Britain their home. This not only transformed the ethnic and cultural diversity of the country, but also transformed the microbiomes of those families travelling across the globe and the microbiomes of the British, whose diet exploded with new flavours.

In 1939 there were just six Indian restaurants in the UK; today there are approximately 12,000, although many also serve foods from Pakistan, Bengal, Bangladesh, Nepal and Sri Lanka. Because of this cultural exchange, chicken tikka masala is now one of Britain's most popular dishes. It consists of marinated boneless chicken cooked in a tandoor and then served in a subtly spiced tomato-cream sauce. Its origin is debated, but one version states that it was invented in the 1970s by a British chef from Pakistan who owned a restaurant in Glasgow. The chef was Ali Ahmed Aslam, and the legend states that one day he was asked by a customer to improve on the dry chicken he was serving. At the time he was suffering from a gastric ulcer, very probably caused by *Helicobacter pylori*, and he had been placed on a liquid diet by his doctor. Taking this as his inspiration, Mr Aslam added a little tomato soup and a sprinkling of spices and, *voilà*, an entire culinary culture was born.

This means that a Pakistani immigrant to Scotland, suffering from a chronic disease caused by bacteria, used chickens fed on antibiotics to cook a meal with traditional Indian spices discovered in Central America, to change the microbiome of an entire nation.

Our diet is now global, which creates unique challenges for the microbiome. Despite stringent food-hygiene laws, our food is not always sterile, and as Élie Metchnikoff feared (see page 124), some of our raw foods are in fact alive. Since the 1970s there's been an exponential global growth in the consumption of sushi – understandable, as nigiri and temaki are delicious. The expansion has been driven in part through the growth of high-street chains and cheaper fish. Yet it has led to 283-fold increases in the abundance of the *Anisakis* nematode found in humans over the same timeframe.

The bacteria within our food share genes for their metabolism and processing in the same way they share genes for antibiotic resistance. Sometimes this is for their benefit, and sometimes for ours. Once again, sushi consumers provide an example of this process. Some Japanese sushi-eaters are only able to metabolize a certain type of seaweed because of a bacterium called *Bacteroides plebeius*. This bug contains an enzyme that breaks down a complex sugar called agarose, which is found in red seaweed. The bacteria is only present in individuals with Japanese ancestry, and it can only perform the task of breaking down agarose because other marine bacteria have transferred the genes that code for this enzyme. If our foods get too sterile, we might not be able to digest them. This also means that our appetite for a global menu is distributing microbial genes for its metabolism across the world.

We are also leveraging the microbiome to understand how

different global diets promote health. Randomized control trials have now demonstrated that when compared to a low-fat diet, a Mediterranean strategy lowers the risk of cardiovascular disease by 30 per cent after five years. Multiple analyses have also identified that this diet changes the gut microbiome, with increases in the abundances of symbionts such as *Faecalibacterium prausnitzii* and *Roseburia* and a reduction in the abundance of *Ruminococcus gnavus*, *Collinsella aerofaciens* and *R. torques*. More recent data suggest that people with gut microbiomes depleted of *Prevotella copri* may particularly benefit from this strategy.

None of these things, however, have stopped the adoption of a hyperglobalized diet. This is food that is fast, homogeneous and served in enormous portions – and it has caused terrible harm to gut microbes and their humans.

Junk food: cheeseburger and chips

We are now eating more than at any other time in our history. The average American's total caloric intake has increased from 2,109 calories in 1970 to 2,568 calories in 2010, and the average meal consumed today is almost four times the size of a typical meal in the 1950s. We have also radically changed how we eat. Much of the world's urban population consumes their diet from ready meals or from takeaways ordered through apps that are seamlessly woven into our digital lives. The global food-delivery market is now worth more than US $150 billion; it tripled between 2017 and 2020, and it doubled again during the COVID-19 pandemic as people dined in, too afraid to eat out. Every year the Uber Eats company releases an Orwellian 'Cravings Report'. In 2021 it proudly reported that orders for cheese fries increased by more than 1,234 per cent during the pandemic.[3] Requests for 'no egg', 'no jalapeños', 'no cilantro', 'no cucumber' and 'no vegetables' also increased dramatically.

The change in the way we eat has transformed our microbiome. Switching from a low-fat, plant-based diet to a high-fat, high-sugar diet with lots of fructose (the typical Western diet) has a dramatic impact on the structure and function of the gut microbiome. When this change in diet is trialled in mice, within a single day their gut bacteria switch on genes that produce metabolites associated with poor gut health. If you are sick, this relationship is even more important. Junk-food diets that are high in fat reduce antibiotic effectiveness, while a high-fibre diet improves the ability of antibiotics to reduce pathogen loads and reduces the number of antibiotic resistance genes in the gut.[4]

Not only have some species bloomed on this ready supply of protein, fat and sugar, but they have mutated and altered the tone and pitch of their conversation with our immune system. To understand how and why, it's helpful to consider how the microbiome *in general* deals with individual components of our diet, because this mirrors how we consider many of the health claims that are made about the foods we eat. The average meal of course varies, depending on whether you are dining in London, Delhi, Hanoi or Shanghai (more than 3.5 billion people depend on rice for their primary source of carb). However, let's look at a typical plate (or box) of food that is commonly ordered in most cultures with an Americanized diet: a delicious cheeseburger and chips (or fries, depending on where you are from) with pickle. Through this one example, we can look at how the microbiome processes both the macro- and micronutrient food groups to make sense of what is going on.

The burger or patty – protein

Although most protein is broken down and absorbed in the small bowel, some does make it to the left side of the colon, where bacterial fermenters such as *Clostridium*, *Fusobacterium*, *Bacteroides*, *Actinomyces*, *Propionibacterium* and *Peptostreptococci* break it down with a family of enzymes called proteases. Some of the products of this process serve as important energy sources for the gut (for example, short- and medium-chain fatty acids), but many of the by-products (such as carbon dioxide, ammonia, phenols, indoles and sulphides) are also toxic.

The abundance of bacteria that make some of these toxic compounds, such as those that produce sulphides, vary between ethnic subgroups. This may also explain why colon-cancer risk varies in meat-eaters between these groups. For example, African Americans have greater abundances of sulfidogenic bacteria, *Bilophila wadsworthia* and *Pyramidobacter* spp., than non-Hispanic Caucasians, regardless of their colon-cancer disease status. These abundances strongly correlate with the components of a diet high in fat and animal protein, and negatively with the servings of dairy and calcium.

One reason to eat meat is that it allows us to put on muscle. The branched-chain amino acids (BCAAs) leucine, isoleucine and valine are particularly important nutritional components of animal protein, although you'll find the same components in dairy and legumes, and the body also manufactures them. They allow us to put on muscle because they regulate a master switch in cells called mTOR (mammalian/mechanistic target of rapamycin) that controls how cells grow and die. In the general population, when mTOR is switched on by high concentrations of BCAA, it causes obesity, insulin resistance and inflammation, and elevated levels of BCAAs in the blood are also a risk factor for cancer. Gut-microbiota metabolism of protein contributes to BCAA synthesis, uptake and degradation, and thus to its circulating levels. If you have lots of *Faecalibacterium prausnitzii* in your gut, then this may be a good thing because it controls transporter genes for BCAAs, lowering the amount of these amino acids that circulate throughout your body and reducing the amount of insulin resistance you experience.[5]

Bodybuilders hack this chemistry by taking leucine supplements to put on muscle, because it activates the mTOR cascade. To be clear, bodybuilders do not have healthy guts, and I strongly advise against this practice because if you change the mTOR pathway, you also change the microbiome – and not in a good way.[6] If nothing else, extremely high-protein diets make bacteria more aggressive, and collectively these diet-induced changes drive inflammation in the gut.

Another important role played by bacteria in meat consumption is the metabolism of choline (an essential nutrient) and carnitine (responsible for energy production). This produces Trimethylamine (TMA), which is used as a carbon source by bacteria in the gut, which convert this into methane, which most of the time is passed (sometimes) discreetly as gas. However, TMA also travels to the liver, where it is turned into Trimethylamine N-oxide (TMAO), and it is the TMAO that causes the harm: it activates inflammatory pathways that deposit fat in our arteries and is attributed to changes in cholesterol levels and bile-acid metabolism. TMAO is associated with heart disease, cardiovascular disease, stroke, liver disease and bowel cancer. The conversion from benign TMA to harmful TMAO cannot happen without bacteria.

The body will metabolize proteins from beef, casein, whey (milk proteins) and soy protein differently, and the microbiome is also in part responsible for this. It's not a simple case of 'animal protein is bad'. It is the quantity and quality of animal and vegetable protein, in the context of fat consumption, and the metabolic consequences of both are determined by your gut microbiota. Processed meats get a deservedly bad press, because they contain a lot of fat.

Animal fats – lipids

When we eat a gloriously sloppy cheeseburger, it is the liver that has the most work to do. For a start, it must produce bile, a fluid that is green in colour and made from cholesterol. One of its many jobs is to act as a detergent, emulsifying fat so that human enzymes called lipases can break down fat for its absorption. Bacteria are critical mediators of bile metabolism, because they modify its chemical structure in the gut to make what are known as secondary bile acids. These bioconversions change the cellular signalling properties of bile acids, which are vital for controlling our metabolism, immune system and even our brain function and appetite. Bile acids also regulate gut microbial composition, both directly and indirectly, because they modify the innate immune response in the intestine. Too much bile is a bad thing, as it is pro-inflammatory and carcinogenic.

Eventually absorbed fats are passed to the liver for processing. Experiments in germ-free animals have shown that gut bacteria also play an important part in determining how the liver metabolizes lipids and cholesterol. High-fat diets change faecal microbial diversity and reduce the abundances of important symbionts, such as *Akkermansia muciniphila*, Christensenellaceae and Tenericutes, which protect the gut barrier, reduce inflammation and increase the production of 'good cholesterol' called high-density lipoproteins (HDL). A high-fat diet also reduces the abundance of small intestinal *Lactobacillus* species (for instance, *L. gasseri*), impairing our ability to use healthy polyunsaturated fatty acids, known as 'PUFAs'.

Humans are unable to make two important PUFAs, and without them we would not be able to live. Omega-3 (linoleic acid) and omega-6 (α-linolenic acid) are used to make a class of lipids called eicosanoids that are important for our immune system. We get these fats exclusively from our diet, and those derived from n-6 PUFA are generally pro-inflammatory, while eicosanoids derived from n-3 PUFA are anti-inflammatory. The problem is that we can't get the balance of these two essential fats in our diets quite right. The ratio of n-6 to n-3 omega PUFAs should be about 1–4:1, but since the 1970s this has climbed to almost 20:1 in Western diets, due in large part to the increased consumption of vegetable oils. Omega-3 can be found in chia and flax seeds, walnuts, cod-liver oil, mackerel and salmon and we really need to eat more of these (sustainable) foods.

You won't be surprised to hear that bacteria can make PUFAs in the gut, and this plays an important part in determining our health. For example, experiments in mice show that they can use PUFAs to become resistant to the bad health consequences of high-fat diet-induced obesity. In humans, various bacteria are also able to produce PUFA-derived intermediate metabolites that have a health benefit. For example, several species such as *Propionibacterium*, *Lactobacillus* and *Bifidobacterium* can produce linoleic acid, which improves insulin sensitivity and decreases atherosclerosis risk. Lactic-acid bacteria like *Lactobacillus plantarum* in the gut are also able to detoxify mono- and polyunsaturated dietary fatty acids that we consume as part of our two-for-one, extra-value meal.

The microbiome also influences our blood cholesterol by influencing how much of it is excreted in our bowel motions. The colon receives about 1g of cholesterol a day, largely through the secretion of bile. Intestinal microbiota can convert this cholesterol into an inert compound called 'coprostanol', which is then eliminated in the faeces without being reabsorbed. This enzyme has only recently been defined; researchers correlated the presence or absence of bacteria that have these enzymes with blood cholesterol levels collected from patients in China, the Netherlands and the US. They observed that people who carry the gene for this enzyme in their microbiome had 55–75 per cent less cholesterol in their stool than those without, which means they were excreting more, and less was getting into the blood.[7] This effectively means that we can engineer the microbiome to reduce cholesterol, and many probiotics are now being produced for this exact purpose, which you will be able to eat with your hamburger.

The oil you cook your burger in will make a big difference to your microbiome. Olive-oil production is also dependent on the colonization of microorganisms for its physicochemical characteristics. Extra-virgin olive oil contains monounsaturated fatty acids and bioactive phenolic compounds that, collectively, are anti-inflammatory and possess important antioxidant activities.

The bun and fries – carbohydrate

Globally, most of the carbohydrate we consume comes in the form of cereals that are used to make the bun of our burger. Of the seventeen primary cereals, rice, wheat and maize (corn) provide more than 42 per cent of all calories consumed by the entire human population. There is a lot of debate about what constitutes the right amount of carbohydrate to eat, but it is absolutely clear that too much or too little is a bad thing and either is associated with a shorter life expectancy.[8] For the average person, the 'sweet spot' is for carbohydrate to make up about half of our calorific intake, because it serves as the basis for multiple biosynthetic pathways that our body really needs to maintain health. However, consuming the right type of carbohydrate is key.

Captain T L Cleave, a naval physician from St Mary's Hospital, spotted the approaching sugar tsunami more than fifty years ago when he published his book *The Saccharine Disease* in 1974. He had been working on a series of studies investigating the blood-glucose response to carbohydrate digestion. Campbell recruited healthy volunteers and fed them 50g of carbohydrate from white or wholemeal bread, apples, maize starch, sucrose and glucose. He ranked the various carbohydrates according to how quickly the participants' blood glucose rose over time, noting that sucrose was not at the top of the list. This concept was eventually developed as the glycaemic index by scientists in Toronto, and is now used all over the world by dieticians, doctors, diabetics and athletes to plan their meals. Diets with a high glycaemic index are more likely to cause type II diabetes. A French fry has a high glycaemic index of seventy-five, and all of that fried carbohydrate will send your sugar rising.

As a rule, refined sugars do not need to be broken down to be absorbed. After a sweet dessert, for example, only about 5–30 per cent of the sugars or sweeteners will reach the large intestine, with the majority being absorbed in the small bowel. The glycaemic index is not very accurate, because it excludes the metabolism of carbohydrates by the microbiome. An Israeli research group studied a cohort of about 800 individuals and 46,898 meals, as they wanted to understand the impact of the microbiome on the postprandial (after a meal) glycaemic responses to several food types. To achieve this goal, they created standardized meals where the glycaemic index and the quantity of calories and carbohydrates were precisely known. Each participant was then asked to meticulously log their daily activities, meals, habits and medications for a period of one week. Each also provided blood tests and a stool sample, so that the metagenomic content of the microbiome could be analysed. From this the researchers were able to build a machine-learning algorithm that incorporated all these measurements, to predict postprandial glycaemic responses with much greater accuracy than carbohydrate-counting or calorie-counting. This in turn enabled them to design personalized meals through which participants were able to precisely control their blood-sugar response to a meal.

Part of the reason this blood-sugar sense magic is possible is that the human genome encodes only a small number of digestive enzymes that break down sucrose, lactose and starch. Instead we rely on a large and diverse group of specialized enzymes encoded by the gut microbiome to metabolize many of the complex carbohydrates we consume. The diversity of these enzymes is immense; the bacterium *Bacteroides thetaiotaomicron*, for example, possesses 260 glycoside hydrolases in its genome alone. Because carbohydrates are involved in a wide range of biological functions, and because the diet of the holobiont has evolved so greatly, genes encoding these enzymes have been subjected to multiple divergent and convergent evolutionary events. We now rely on these evolutionary links to metabolize our cheeseburger.

Sugar itself has a massive impact on the ecology of the gut. A high intake of refined sugars increases the relative abundance of the pro-inflammatory Proteobacteria in the colon, while simultaneously decreasing the abundance of Bacteroidetes, which promotes gut-barrier function. Fructose, which is commonly added to condiments like tomato ketchup, also increases phage production in *Lactobacillus reuteri* through a stress–response pathway, which in turn changes microbiome composition.[9]

Microbes in the gut rapidly change their metabolism to take advantage of sugar availability. This means that different bacteria from the same species, such as *Bacteroides fragilis*, can develop different ways to metabolize or transport sugars in the same gut. Bacteria will also use these sugars to make tools necessary for their survival, such as biofilms and flagellar structures that enable their mobility and their pathogenicity, further exacerbating inflammation within the bowel. Not only does this drive inflammation in the bowel, but more microbial genes available for sugar metabolism and transport mean there is a higher risk of type II diabetes and obesity.[10]

Not only have humans co-evolved with bacteria that help us

consume sugar, but we've also evolved dedicated brain circuits to seek, recognize and motivate its consumption. We're now starting to understand how this works. There's a set of neurones in the part of the brainstem where the vagus nerve originates that works via the gut–brain axis to create a preference for sugar.[11] These neurones are activated as soon as sugar (but not sweeteners) hits the gut. In mice, if the vagus-nerve pathway is artificially inactivated, their preference for sugar is shut off. This represents another opportunity for our evolutional microbial partners to regulate our food-seeking behaviours via the gut–brain axis. Collectively, our microbiome makes some of us very susceptible to the taste and metabolic consequences of carbohydrate consumption.

The problem is that all of these circuits are being overwhelmed, as almost 13 per cent of our carbohydrate income in the USA now comes from added sugar. This is a huge amount when you consider that the recommended amount of free (or refined) sugar suggests that we should not exceed 5 per cent of our total dietary energy.[12] We are completely addicted to this bountiful white gold, and artificial sweeteners may not be much better. Early data on artificial sweeteners, such as saccharin, suggested they could also induce a glucose-intolerance-promoting microbiome.[13] Although these findings are controversial, it is still probably a good idea to think about how many we put in our coffee.

The lettuce leaf – fibre

Fibre is a genuine superfood and we should be eating A LOT more of it than a single limp leaf of iceberg lettuce that has been reluctantly forced into a mass-produced hamburger.

In the 1970s a British surgeon called Denis Burkitt, working at the Mulago Hospital in Kampala, Uganda, wondered why his patients didn't suffer from many of the diseases he saw in the West. His work would define the importance of fibre to human health. The medical profession faced an uphill battle when it came to spreading his message, however – not helped by fibre's unfair reputation as being only a little less tasty than cardboard. A bigger problem is that we've never been able to agree on what, precisely, fibre is, and how it benefits our health. That's because the term 'fibre' represents a superfamily of plant-based molecules with a spectrum of physical and chemical properties. Essentially, to be a fibre, this food has got to get to the colon and has to have a specific chemical formulation that means only bacteria can break it down.

If you can increase your intake of fibre by just 7g per day (about 3.5 apples, two cups of peas or 3.5 tablespoons of bran) you will statistically reduce your risk of cardiovascular disease, stroke, colorectal cancer, rectal cancer and diabetes. You really need at *least* 30g of fibre a day in total, and you need to have this consistently.

What exactly makes this wonder-food work? Broadly, fibres come in two forms: non-starch polysaccharides that originate in the cell walls of plants (cellulose, hemicellulose, lignin or pectin, β-glucan and galactomannan); and resistant starch, a plant storage carbohydrate present in cereals, legumes, rhizomes, roots and tubers. Potatoes contain fibre, but if you are going to have chips, opt for sweet-potato fries, which have the highest fibre content.

Some starches are very resistant to digestion and their health benefits vary greatly, based on this fact. Fibres used to be described as being soluble or insoluble. But this isn't helpful, as many vegetables and fruit contain both, and the health benefits of fibre aren't necessarily determined by solubility. Each type of fibre also has a different pattern of viscosity, which means it pulls water into it and forms a gel. This means that some fibres such as psyllium or β-glucans bulk the stools, increase stool weight and slow their transit (there is literally more to shift along the gut), which changes how we absorb iron and vitamins, including vitamin D. Fibre also changes how the small bowel absorbs bile acids, which may explain its cholesterol-lowering properties. Supplementing our diets with wheat bran, on the other hand, increases stool frequency and changes gut motility. But one of the most important health benefits of fibre comes from its fermentation by bacteria.

Some species of bacteria, such as *Ruminococcus bromii*, have highly specialized starch-degrading enzymes, while others are generalists.

In practice, microbes metabolize fibre through a mutualistic network, distributing the work through a shared number of enzymes and fibre metabolism. This is known as 'cross-feeding' and it allows the gut to break down fibres as they pass along the long tube. The acidity (pH) of the gut changes with its metabolism as it progresses along, and this in turn regulates bacterial populations. Farting is a good example of cross-feeding. Hydrogen is produced by fermentative gut bacteria and consumed by three different microbial groups – sulphate-reducing bacteria, acetogens and methanogenic archaea – that work to produce hydrogen sulphide (smelly), hydrogen (blameless) or methane (flammable). This is why many people can't tolerate a high-fibre diet. If you're moving to a high-fibre diet (as you should be), do it slowly over many weeks to give your bacteria a chance to catch up.

Professor Stephen O'Keefe of Pittsburgh University has built on Burkitt's work by investigating the effects of fibre on the microbiome in ethnic cohorts at extremely high and exceptionally low risk of bowel cancer.[14] African American men have some of the highest rates of colorectal cancer in the world. There are, of course, multiple factors behind this, including socioeconomic disadvantage and reduced access to healthcare. Meanwhile, rural South Africans have some of the lowest rates of bowel cancer in any population. O'Keefe asked a group of people from each ethnic cohort to swap their diets over for two weeks. This is an extremely hard thing to do. The Americans struggled with the high-fibre, low-fat, low-protein diet they were given every day for two weeks, and the Africans did not like all that meat and fat.

After just two weeks of dietary exchange, the results were spectacular. The researchers didn't see large changes in the abundance of bacteria, but they did observe dramatic changes in the way the different elements of the microbiome work together to metabolize food. The Americans shifted within two weeks from a gut engine designed to metabolize meat to one that ferments fibre. With that came dramatic increases in the sorts of metabolites these bacteria produce, and which are important for gut health. For example, they observed a rise in the concentrations of faecal short-chain fatty acids (these metabolites are anti-carcinogenic) and reductions in bile-salt excretion (which is carcinogenic). This caused levels of inflammation in the gut to fall, as did the risk of cancer (measured by biopsies from the gut). The Africans experienced reciprocal changes, in that their bile-acid excretion rose by 400 per cent and their short-chain fatty-acid excretion fell dramatically. Interestingly, none of the African participants in the study had colonic polyps, which are precursor lesions of cancer in the colon, whereas 40 per cent of the Americans did. This work is important as it shows that cancer risk can be dramatically changed very quickly with the right dietary intervention, and it is an excellent reason to start eating veggie burgers.

But changing to this high-fibre diet isn't possible for all Western guts; some people's microbiomes simply can't tolerate it, at least not if the change happens too quickly. And fibre alone doesn't tell the whole story.

The tomato and onion – polyphenols

I don't trust a cheeseburger without a tomato, because it adds colour and flavour. This is also where the mighty polyphenols live.

From a nutritional perspective, a tomato is made up of water (95 per cent) and the rest is predominantly carbohydrate, in the form of glucose, and small volumes of fructose. However, it also contains a little fibre and traces of vitamins, minerals and lycopene, which is a type of plant pigment known as a carotenoid. This gives the tomato its red colour, and lycopene has established anti-carcinogenic properties, particularly in the prostate.

Polyphenols are chemical compounds whose job is to defend plants from ultraviolet radiation or pathogens. Humans don't produce polyphenols, but they improve human health and reduce the risk of chronic inflammatory disease and prevent cancers. Polyphenols are found in fruits such as grapes, apples, pears, cherries and berries. Typically, a cup of tea or coffee contains about 100 mg of polyphenols, as do cereals, dry legumes and chocolate.

Historically, the health benefits of polyphenols have been thought to come from their antioxidant properties, but microbiome science is changing this view. Much of their health benefit actually derives from the colonic microbiome gorging on them with the greedy enzymes that love nothing more than breaking down polyphenols. This has lots of benefits; for example, it promotes the production of short-chain fatty acids that fuel the colon and regulate brain function. Polyphenols also promote the secretion of mucus in the gut through a direct effect on the mucin cells that line the gut. They promote the growth of certain bacteria, such as *Akkermansia muciniphila*, which stimulates mucin production, and they inhibit the growth of mucus-eroding bacteria. Finally, polyphenols directly interact with both the innate and adaptive immune systems to minimize inflammation. Beyond promoting the growth of symbionts, they also inhibit the growth of pathogenic bacteria.[15]

The pickle – fermented foods

A pickle contains fibre, vitamins and lots of water and possesses very little protein and fat. It is the vinegar, however, that gives the tang we need in our cheeseburger, and it is the fermentation of vinegar that brings the health value. Vinegar production is an ancient practice that relies completely on its own microbiome, and it was Pasteur who first explained what it is and how it works, concluding that fermentation is the consequence of 'life without air'.

In 2020, during the first lockdown of the pandemic, many of us turned into artisanal food manufacturers, carefully nurturing our yeast starters or SCOBYs (symbiotic cultures of bacteria and yeast) as we sought to produce our own kefir, kombucha, sauerkraut, tempeh, natto, miso, kimchi or sourdough bread. These typically use fungi, such as *Saccharomyces* for bread, beer and wine; *Aspergillus* for traditional fermented Asian foods, such as sake, soy sauce and miso; and *Penicillium* for cheese and cured or fermented meat. Vegetable and animal fermentations are also determined by plant- and animal-associated microbiota respectively, and these are often consistent across cultures and manufacturing processes. For example, fermentations of cabbage and other green vegetables go by different names in different parts of the world. *Sauerkraut* in Europe and North America, *kimchi* in Korea, *suan cai* in China and *sinki* in Nepal are all variations on the same theme and are relatively consistent in the active microorganisms used to make them.

How fermented diets change the microbiome is now slowly being unravelled. In 2021, a team at Stanford reported on a seventeen-week prospective study where thirty-six healthy individuals were put on either a high-fibre diet or a high-fermented-food diet.[16] There were marked differences in the gut microbiomes between the two groups, and those on a fermented-food diet not only demonstrated a greater microbiota diversity than those on a fibre diet alone, but also had lower levels of circulating inflammatory markers. This means that a combination of high-fibre and fermented foods could be a better way to optimize the benefit of a high-fibre diet alone. My own stance is that fermented foods have a positive impact on gut health, and I recommend them to my patients.

Soybean products, such as the pungent *cheonggukjang* (Japanese *natto*), *doenjang* (soy paste), *ganjang* (soy sauce) and *douchi* (Chinese soy paste), are widely consumed in East Asian countries and are made both at home and in factories. Soybeans are typically fermented with bacteria and fungi, and the soybean produces its own unique source of bioactive molecules, such as isoflavones, soya-saponins, lignans, cinnamic-acid derivatives, terpenes and sterols. The reported health benefits for these include a lower risk of neurocognitive decline and improved brain health.[17]

Soy can be consumed in lots of different forms, and it's increasingly drunk as milk. Soy polyphenols known as isoflavones give rise to phytoestrogens such as equol, which are endocrinologically active. Japanese and Korean menopausal women who consume copious amounts of soy milk are less likely to get hot flushes, but both men and women who consume it regularly have less risk of chronic diseases in general. Men reduce their risk of prostate cancer, for instance. Interestingly, we are not all born *equol*; some of us can metabolize equol and some of us can't, depending on the bacteria in our gut. More than half of Japanese and Korean people are equol-producers compared to just one in seven in the West, although Western vegetarians are four times more likely to be equol-producers than omnivores, possibly because of their increased fibre consumption.[18] But it seems the only way to become an equol-producer is to consume a lot of soy milk.[19]

The known unknowns – additives

Feeding eight billion people worldwide means that we've had to develop safe, nutritious food that tastes better and lasts longer. We add artificial sweeteners, sugar alcohols, emulsifiers, food colourants, flavour enhancers, thickeners, anti-caking agents and preservatives. We know very little, really, about their relationship with the microbiome, but we do know the relationship exists: food additives have been associated with inflammatory bowel disease,[20] food allergies and irritable bowel syndrome.

Emulsifiers – one type of food additive – are made from plant, animal and synthetic sources and are commonly added to processed foods such as mayonnaise, ice cream and baked goods to create a smooth texture, prevent separation and extend their shelf life. The FDA regards them as generally safe. Animal studies suggest that they can indeed influence gut-barrier function and reduce intestinal microbiota diversity while increasing the number of mucolytic microbes, such as *Ruminococcus gnavus*, that break down the barrier. The intestinal microbiota appears to be particularly vulnerable to the emulsifiers CMC and P80, and experimental data suggest that they may increase the abundances of pro-inflammatory bacteria.[21] This topic remains highly contentious, as emulsifiers make food cheap and accessible, and I do not want to demonize ice cream or my children will not forgive me.

Part of the challenge is the variation in how we all metabolize additives. Scientists from the University of Michigan studied how a thickening agent called xanthan gum (XG), a polysaccharide made by the plant bacterium *Xanthomonas campestris*, influences the gut microbiome. They observed that some people in industrialized countries can metabolize this additive and others can't. The difference comes down to the presence of a single uncultured bacterium in the family Ruminococcaceae. It gets more intriguing because some individuals also had a microbe called *Bacteroides intestinalis* that can't consume the gum, but can grow on sugar products generated by the Ruminococcaceae.[22] The true health implications of this symbiotic transformation aren't known, but additive–microbiome interactions are likely to be far more complex than is currently understood.

But what if you like your hamburger rare or well done? As far as the microbiome is concerned, it's all in the cooking.

Cooking for the microbiome

How we cook and prepare our food makes a big difference to the gut microbiome. It kills pathogens that may be present in our food, but it may also destroy some of the natural antibiotic properties of raw foods and their more helpful symbionts.

Because the process of cooking involves mixing, beating and heating up ingredients, it changes their physical properties. This will have a major impact on the ability of the microbiome to access and metabolize components of our diet once it is consumed. How you cook your vegetables is of particular importance because resistant starches are locked into the food matrix, and cooking turns resistant starch into gelatine, making it more soluble so that it can be digested. The University of San Francisco demonstrated the consequences of cooking starches on the microbiome in a study where mice were fed both cooked and raw organic lean beef or potatoes.[23] Cooking the beef didn't seem to make a substantial change to gut microbial communities. But cooking made a remarkable difference for those animals given a potato diet, where overall gut microbiome diversity decreased and the relative proportion of Bacteroidetes – a phylum important in carbohydrate breakdown – increased. The decreased diversity could in part be explained by the fact that the animals were given a less varied diet than in their normal feed.

However, the researchers also noticed that raw diets caused mice to lose weight.

When similar cooked and raw diets were tested in human volunteers, the same effect on the microbiome was observed. Cooking the food containing resistant starch may have allowed the small bowel to soak up more calories, as it wouldn't have had to work so hard to break it down. This is a big deal for the microbiome canteen: it means there may be less to go around in the colon and, as a result, diversity falls.

Shakes and liquid drinks have also come into fashion and these have a similar effect. Several companies now market ready-made powders or liquids that completely remove the need to cook, and they claim to offer a complete set of nutritional requirements for consumption at your Silicon Valley desk. But the content of this liquid diet may be less important than the fact that it is liquid, as the size of the fibre particle has a significant effect on its absorption, viscosity and rate of fermentation (which can actually increase).

In clinical practice, 'exclusive enteral nutrition' – food administered strictly through a liquid formula – is used in treating children with the inflammatory bowel disease, Crohn's. This liquid-diet formulation is so effective that it's used instead of steroids. And yet, despite many trials, the exact reason for its efficacy is not known. A team at the University of Amsterdam Medical Centre studied the role of the gut microbiome in mediating this effect, as part of a trial in forty-three children. Amazingly, both the microbiota and their metabolites differed between those children who responded to the treatment and those who did not. So it was possible to predict who would get better, based on the metabolic functions of the bacteria present in the gut.[24] The enteral feed dramatically reduced microbial diversity and amino acids, trimethylamine and cadaverine towards control levels, and it reduced the metabolism of bile acids.

Nutritional shakes probably have a similar effect, which is why they promote weight loss. In other words, they change the function of the microbiome, alter bile-acid metabolism and the production of secondary bacterial metabolites and reduce inflammation. But most of us *don't* have Crohn's disease. Our symbiotic bacteria need solid food to break down so that they can feed themselves across networks, and they need both raw and cooked food. This is why a high-fibre diet is so important for our health. We also need to socialize while we cook and eat, because eating together is fun, it's good for our mental health and for sharing microbes. I don't want to live in a world where fine dining means drinking out of a plastic bottle while posting photos on Instagram about how I am living my best life, while I slowly die inside of loneliness.

We should all take responsibility for preparing and cooking our own food if we want to get the best out of it, for our health and our microbiome.

A drink at the microbiome bar

When we eat, we often drink alcohol – and we drink a lot. According to the World Health Organization, the citizens of the Czech Republic consumed the most alcohol per capita in 2019, with individuals consuming about 14.26 litres of pure ethanol over the year: that must have been some party! The United States is a little less drunk, but it still had an annual consumption per person of 9.97 litres.

There's no doubt that alcohol consumption changes the microbiome, and the reasons for this are multiple. First, alcohol contains a lot of calories, and this varies depending on your tipple. For example, one pint of lager (550 ml) has 250 kcals; a standard glass of wine (175 ml) has 130 kcals; and an alcopop bottle (275 ml) has 198 kcals. Second, there is a wide variation in the precise carbohydrate and sugar content of alcoholic drinks, with alcopops, ciders and beers containing the most. The quantity consumed, the timing of that consumption (for instance, adolescence vs middle age), the sex of the person drinking it and, most obviously, the amount consumed all play a role in defining its health consequences, the level of inebriation and its impact on the gut microbiome. Finally, wine-makers and breweries know all about the microbiome, because fermentation is core to the manufacture of alcohol and so the by-products of bacterial and fungal metabolism will vary subtly between brands, which all use their own secret recipe.

Having said all this, a study of 916 women in the UK demonstrated that red-wine drinkers have a greater diversity in their gut microbiome compared to people who drink white wine, beer, cider or spirits.[25] The more frequently wine was drunk, the greater the change, but even an occasional glass had an effect on the gut microbiome. Red-wine drinkers have more of the bacteria genus *Barnesiella* in their gut, and changes in the red-wine microbiome correlated with a low Body Mass Index (BMI). This could of course reflect the fact that red wine contains less carbohydrate and sugar.

However, the production of red wine also includes the skin of the grapes, which means it possesses a tenfold higher polyphenolic content than white wine. This exerts its beneficial effect through the colonic microbiome. Resveratrol – a type of polyphenol in red wine – might contribute to health and ageing benefits because of its anti-inflammatory, antioxidant, antiviral, antifungal and anti-carcinogenic properties. These observations are in keeping with many epidemiological studies on ageing suggesting that an occasional glass of red wine is good for the health (thank God).

Binge-drinking lowers the diversity of the gut in both humans and animals, although this is likely to be confounded because a night on the sauce is more commonly associated with poor food choices. Alcohol addiction is also more likely to lead to malnutrition. Despite this limitation, individuals with alcohol-dependency develop an inflamed gut, and bacterial toxins are able to filter through the leaking gut.[26] The faecal microbiome of baboons was studied in an experiment where one group was given regular alcohol consumption for more than twelve years and another was given alcohol over a shorter term (three years). The long-term cohort exhibited a loss of diversity compared to the short-term group, and the growth of pathogens such as *Streptococcus*. Long-term drinkers also showed signs of intestinal inflammation, although the diversity of the microbiome was relatively stable from day to day and there were few differences between animals drinking for a short time and the controls. However, despite a stable microbial ecology, the researchers found much larger changes in the metabolic functions of the microbiome as animals transitioned from drinking to abstinence.[27] This means that although the ecology of the microbiome does not change greatly with abstinence after prolonged periods of alcohol consumption, the function of the gut microbiome will very much change for the better.

The microbiome influences the liver's ability to detoxify alcohol because, just like paracetamol, it influences the availability of sulphur needed by the liver's enzymes to break it down. Because the gut microbiome contributes directly to liver injury from excess alcohol consumption, it can also be targeted as a treatment to reverse this effect. For example, the bacterium *Enterococcus faecalis* found in the gut microbiome of a drinker creates a specific toxin that damages the liver, called cytolysin, which causes inflammation known as alcoholic hepatitis. When faeces from alcoholic patients with high levels of this bacterium was transplanted into germ-free mice, phage isolated from sewage water could be targeted to *E. faecalis*, which prevented the toxin from damaging the liver.[28] This does not mean that you should drink sewage water with your wine, but rather that the microbiome can be engineered to prevent liver injury in those addicted to alcohol.

Bacteria in the gut can also serve as our own personal brewery. In 2016, a woman in Upstate New York was driving her husband home from dinner when one of her car tyres burst. When the police stopped to help, she was routinely breathalysed and found to be twice over the legal limit for alcohol. She was booked for driving under the influence, but claimed that she had barely been drinking. In court, her defence was that her gut was acting as its own microbrewery.[29] This is not as bonkers as it sounds: there are in fact several reported cases of 'alcohol fermentation syndrome', and most of these are the result of yeasts such as *Candida albicans* and *C. krusei*,[30] which ferment sugar in the gut into ethanol.

If you decline an alcoholic drink at the microbiome bar, you may choose a soft drink instead, but take my advice and pause before you order. By the turn of the millennium, soft drinks accounted for almost 9 per cent of overall energy intake in the average American, and today 13.5 million litres are consumed in the UK each year. In 2015, a group from Tufts University in Massachusetts estimated that among all worldwide annual deaths from diabetes and cardiovascular diseases, about 178,000 were attributable to sugary-drink consumption alone,[31] and fructose has a lot to answer for, in part because it also influences gut microbial diversity.

Let's now look at whether this is a case of 'everything in moderation' or whether we need a more fundamental change.

A change of menu

The Western and urbanized diet is homogeneous, colourless, bland and dripping in sucrose and fructose and high-glycaemic carbohydrates, trans-fats, omega-6 PUFAs and salts. And we're told that we are 'lovin' it'. Except that we are not – and neither is our microbiome. When Cristiano Ronaldo removed a Coca-Cola bottle from his post-match interview on live TV during a Euro 2020 press conference and replaced it with water, he wiped US $4 billion off the company's share price. Coca-Cola needn't have worried, though. As Uber Eats brazenly reminds us, we are chemically addicted to fat and sugar, and the share price soon bounced back.

Our relationship with food and drink in the twenty-first century is a paradox: we've never had so much, and we've never wasted so much. Some of us can't stop eating and need surgery to stop, while others must be fed against their will. In developed countries we can't decide what constitutes a healthy diet, while those in low- and middle-income countries starve to death. Food inequality is a modern-day tragedy, and food security is an impending global crisis. Our elite athletes measure out every component of their diet to the nearest microgram to ensure maximum performance, and cancer sufferers don't know what to eat to help them recover. We don't understand how most foods – let alone microbes – cause chronic disease, and our society is full of people living at nutritional extremes. All of this has become entirely normalized.

These challenges are massive and multifactorial. But the microbiome offers a new perspective from which we can begin to reverse-engineer solutions to some of our greatest nutritional problems. Microbiome science makes it clear that both our environment and our gut need to move to sustainable plant-based diets that emphasize the consumption of fruit, vegetables, nuts, seeds, oils, wholegrains, legumes and beans, with small or moderate amounts of meat, fish, seafood, eggs, dairy and wine. We are going to need fewer fizzy drinks and more red wine – but only in moderation. We must also first learn to meaningfully incorporate the microbiome into the science of gastronomy. This means sustainable cooking that selectively targets the microbiome, not only for our health, but also to heighten the sensory experience and joy of eating. This requires a new 'personalized' measure of our diet, based on an analysis of our food, our microbiome and our genes.

If we get it right, our food could taste incredible. If we get it wrong, we are all going to need a menu for an FMT. *Bon appétit.*

PART THREE
The Metahuman

14. Extreme Phenotypes

A bariatric operating room is the state of the art in weight-loss innovation, and it has transformed the futures of many patients living desperately unhappy and unhealthy lives. Being in a bariatric weight-loss centre is like being on another planet, where everything looks the same, but is three times the size. The chairs in the waiting room are like double beds, and the giant operating-room tables are reinforced so that patients who weigh more than 200 kg can be safely supported while they undergo treatment. The surgical instruments are twice as long as normal, because they must reach through the thick armour of subcutaneous fat that protects the stomach and small bowel. Either the gastric stomach is banded or stapled, to restrict the volume of food it can contain, or the small bowel of the foregut is precisely re-plumbed to create a state of surgically induced malabsorption. Every time I step within this environment, I ask myself: how did we get here?

By the start of the twentieth century the tsunami of obesity was building, seemingly pulling everyone eating a Westernized diet of high fat, high protein and low fibre under its cold, dark waters. In 2016, more than 1.9 billion adults were overweight and 650 million were obese, and by 2017, 8 per cent of global deaths were attributable to obesity.[1] During the pandemic, the American CDC reported that the Body Mass Index (BMI) in those aged between two and nineteen doubled, as our access to social support failed and we did our best to cope.[2] The result has been a pandemic-within-a-pandemic: *'covesity'* has made the literal and figurative obesity burden for our children heavier than ever.

Make no mistake, the obesity pandemic is a global crisis of our own making, and our young and the most deprived members of our society are paying the greatest price for our failure to prevent it. But the precise reasons for this crisis are not as simple as we would all like to think, and I strongly believe that no one chooses obesity.

The thermodynamic theory of obesity states that we're gaining weight because we're eating more, but this doesn't explain why some people are so vulnerable to becoming obese, and why some obese people develop diabetes and others don't. Although there is a heritable component to obesity, our genes only explain a minority of the overall genetic risk, which is about 20 per cent for type II diabetes and less than 5 per cent for BMI. Even the so-called 'fat genes', such as FTO, account for less than 3 per cent of population variance in BMI. The thermodynamic theory is therefore an oversimplification of our biology, which has perpetuated the trope that obesity is a combination of laziness and gluttony. It ensured that obese children everywhere were bullied into adulthood, and that a culture of fat-shaming has pervaded across the same media agencies that advertise high-fat diets to our children.

Obesity is such a pernicious crisis because the common biological pathway that causes its associated health problems is inflammation, and this is the same route through which almost all chronic disease – including cardiovascular disease and cancer – develops. It even influences our susceptibility to acute infections; as the then British Prime Minister Boris Johnson personally discovered, obesity dramatically worsened the outcome of those who contracted COVID-19. Of the 2.5 million COVID-19 deaths reported by the end of February 2021, 2.2 million were in countries where more than half the population was classified as overweight.[3]

The microbiome has been the biological middleman in the rush to hyperglobalization that has pushed humanity to nutritional and metabolic extremes – and it will define our future risk of chronic disease. Understanding how and why the microbiome is so important to the development of obesity is crucial in deciding how we can act upon this crisis.

The obese microbiome

In 2004, the microbiome pioneers Fredrik Bäckhed and Jeff Gordon from the University of Washington published the findings of a study which fundamentally changed how we think about obesity. In this study, microbes were taken from the colons of conventionally reared obese mice and transferred to germ-free mice. This caused a staggering 60 per cent rise in the body fat of the germ-free animal. The scientists showed that obesity affects the diversity of the gut microbiota, and that obese animals had a 50 per cent reduction in the abundance of Bacteroidetes and a proportional increase in Firmicutes. The findings suggested that the community structure of the microbiome could be manipulated either for weight gain or for regulating energy balance in obese individuals. This is because the microbiota promoted the absorption of monosaccharides from the gut and triggered the production of fat in the liver. Later experiments would demonstrate that microbiota don't only influence how fat is stored, but also how it is burned.

In 2013, Jeff Gordon's team published their data to this effect. They studied the microbiome of female identical twin pairs, where one was obese and the other was lean. (If genes were wholly responsible for weight gain, this should be unlikely.) They then transferred their microbiomes into germ-free mice, which were fed low-fat mouse chow as well as diets representing different levels of saturated fat, and fruit and vegetable consumption typical of the US diet. Mice given the faecal transplant from the obese twin became obese, while those from the lean twin did not. Gordon and his team discovered that obesity and the metabolic consequences of obesity were transmissible between individuals of the same species, and that this could even be done between species – for example, between a mouse and a human. Still more remarkably, when mice with the obese twin's microbiota were allowed to live with those containing the lean twin's microbiota, obesity was prevented because specific members of Bacteroidetes were transferred between them. With these seminal studies, it was understood that our Westernized diet had forced our microbiome engine to become highly efficient at harvesting energy. This change was so dramatic that it overwhelmed whatever genetic defence we had against it, and in genetically susceptible individuals the consequences were dramatic. All at once it became possible to imagine that the microbiome could be targeted to prevent or treat obesity.

However, obesity is really a complex disease that is closely associated with a diverse range of metabolic conditions, such as diabetes and cardiovascular disease, and these are variably experienced by overweight people. The metabolic consequences of obesity cannot therefore be explained in their entirety by a single analysis of gut microbial diversity. In reality the microbiome regulates food metabolism, host energy control, fat accumulation and gut-barrier integrity through shared molecular functions. And there is evidence that some strains play a more important part in this process than others. For example, obesity is associated with lower intestinal abundances of mutualistic bacteria such as *Akkermansia muciniphila*, which improves gut-barrier function and insulin sensitivity. The importance of our missing microbes to our obesity susceptibility is also not clearly understood. The role played by the microbiome in the causation of obesity is therefore much more nuanced.

The microbial co-metabolism of our cheeseburger, for example, also influences how we think and feel about the next plate of food that we want to eat, even before it's digested. This is because the gut microbiome works through the gut–brain axis to change our food-seeking behaviour. For example, microbial co-metabolites work directly on the brain, in a region called the hypothalamus that controls our hedonistic response to food through the production of neurotransmitters such as serotonin and GABA. The microbiome has influenced how we became addicted to sweet and fatty foods.

BCAAs (from our burger), short-chain fatty acids (from vegetables) and bile acids (from fatty foods) that are co-produced by our gut bacteria change our appetite centrally within the brain.[4] They can also do this within the gut. Short-chain fatty acids communicate with cells that line the gut to secrete hormones (called PYY, GLP-1) that suppress our sense of hunger, which is another reason to eat vegetables.[5] Our gut bacteria are also able to directly influence starvation signals by interfering with the function of two other important appetite hormones: the brake, called leptin (appetite suppressant), and the accelerator, called ghrelin (hunger hormone), which sends signals to the brain via the vagus nerve.

Synthetic fructose, for example (found in fizzy drinks and junk food), does not cause obesity through a calorific effect. First, it interferes with the way fat cells use energy, and it directly induces resistance to leptin, secreted by our fat cells. It essentially tricks the brain into thinking the body is always hungry, so we keep eating. Leptin resistance is associated with a lower bacterial richness in the gut, and both fructose and glucose also inhibit gut colonization by symbiotic bacteria such as *Bacteroides thetaiotaomicron*, which are more commonly found in lean and healthy individuals. Collectively, all of these things promote inflammation, potentiating the damaging cycle of microbiome manipulation of our appetite. But the most important hormone in this cycle of damage is insulin.

Syndrome X

The hormone that predominantly regulates our blood sugar is called insulin. Type I diabetes is an autoimmune disease that typically presents in young people who are lean and genetically susceptible. Both innate and adaptive immune systems are inappropriately activated and involved in disease progression, which is caused by the destruction of specialist cells within the pancreas called β-cells, which stop secreting insulin. This means lifelong insulin therapy. The gut microbiota of type I diabetic patients and patients at risk for type I diabetes mellitus (DM, the medical name for diabetes) are distinctly different in their structure and function from those of people without this condition. Individuals affected by type I diabetes also have higher rates of intestinal permeability and an inflamed gut. So is the microbiome influencing the risk of type I diabetes in genetically susceptible children, and why is this useful for understanding how the microbiome causes obesity? Once again we can look to divided geographical regions for clues.

Karelia is a staggeringly beautiful wooded region in northern Europe that is politically significant to Finland, Russia and Sweden. It has a turbulent history, and this region is currently divided between the Republic of Karelia in the north-west of Russia and Finland. Children growing up on the Finnish side have an incidence of type I diabetes that is five to six times greater than those growing up on the neighbouring Russian side, and they also have a higher risk of allergies. Nearby Estonia, which sits between Russia and the Gulf of Finland, has also started to experience rising rates of type I diabetes in parallel with its economic development. A recent analysis of gut microbiomes of infants in the region found that *Bacteroides* species appear in low abundances in Russians, but dominate in Finns and Estonians, whose microbiome was also enriched for genes that code for lipopolysaccharides (which can behave as toxins) and others involved in the production of bioactive molecules.[6] These endotoxins, produced by *Bacteroides dorei*, are structurally different from those of *Escherichia coli*, which was the predominant endotoxin in the Russian gut. The Finnish and Estonian gut bacteria had effectively silenced the *E. coli* that would otherwise have educated the immune system, potentially preventing type I diabetes from developing. In other words, the increase in type I diabetes is linked to a loss of the segmented filamentous bacteria that help maintain the gut barrier and regulate inflammation – findings that have been supported by other animal studies.[7] For example, when a raised blood-sugar level is corrected with insulin in mice, there is no change in the microbiome. But when an anti-inflammatory medicine is given, the microbiome changes and the blood-sugar level normalizes.

As we've explored in this book, the microbiome has a large bag of tricks for regulating inflammation in the gut, so there are likely to be other ways in which the gut is involved in causing diabetes. It produces large amounts of anti-inflammatory metabolites (the short-chain fatty acids we've seen elsewhere), for instance, which have also been shown to have a protective effect in early-onset human type I diabetes.[8] It is also probable that the gut's dark matter is playing an important role in this process. For example, viruses from the gut have been found in the cells of the pancreas that produce insulin, and the rate at which these viruses replicate and their pathogenicity have both been linked to autoimmunity. For instance, young children with prolonged exposure over many years to a virus called enterovirus B and reduced exposures to a virus called mastadenovirus C were more likely to develop autoimmune changes in the β-cells of the pancreas, although this did not translate into higher rates of type I DM.[9]

Type II diabetes is very different from type I, in terms of both its causes and its symptoms; by 2030, 643 million people are predicted to live with type II,[10] a chronic disease closely associated with obesity, typically caused by insulin resistance leading to high blood sugar. The different types of fat that we all have, and its distribution around our body, are major determinants of insulin resistance, and our BMI – our weight divided by our height – is not a particularly accurate measure of our metabolic health or of the severity of obesity. In reality, type II diabetes appears at a tipping point, when there is not enough insulin being secreted by the pancreas, and the insulin that is being secreted no longer works in tissues to reduce the blood-sugar levels because it has become resistant. It's possible that similar bacterial functions in the gut that distort the immune system to halt insulin production in type I diabetics are doing something similar, but far less severe, in type II diabetics.

Studies of the microbiome are now starting to provide some insight into type II diabetics. A recent analysis of more than 2,166 Dutch people, which accounted for many of the environmental confounders of the microbiome, confirmed that those with type II diabetes have less microbiome diversity and less butyrate production in the gut, compared to age- and sex-matched non-diabetic people.[11] So why is this happening? Is it all down to diet?

The answer is no. As you will have worked out by now, antibiotics don't only promote growth in animals, they do the same thing in humans. An estimated 80 per cent of all antibiotics consumed in the US are used in the feed of animals intended for human consumption, and these medicines infuse our world. Epidemiological data have shown that the duration of antibiotic use closely correlates with type II diabetes risk, particularly in women.[12] In mice, antibiotics reprogramme the cells of the gut so that they change their primary fuel source from short-chain fatty acids to sugar.[13] Antibiotics change food-seeking behaviour, the way the pancreas secretes its hormones, the metabolism of bile and every other metabolic pathway right along the gut's great chemical superhighway. It's why they have also been successfully trialled as a treatment for malnutrition.[14] In many ways we are just like battery hens, stuffed full of antibiotic growth products and carbohydrate, unable to imagine a world beyond the slaughterhouse.

The timing of gut inflammation is extremely important when it comes to causing obesity and diabetes, and this may happen much earlier than previously thought, through the process of orchestral signalling. For example, a loss of short-chain fatty-acid-producing bacteria in the guts of pregnant mothers on a high-fat diet has important implications for their developing babies. Not only do short-chain fatty-acid metabolites suppress insulin signalling and reduce fat deposition, which is protective in the mother, but they are also able to pass across the placenta and into the embryo, and a particular metabolite called propionate is able to programme insulin signalling and the development of the sympathetic nervous system in the unborn baby.[15] The good news is that a high-fibre diet was enough to correct this in an animal model, although it also means that the programming of our metabolism in very early life is extremely important for our long-term health.

The implications of the low-level inflammatory process in the gut in obesity and type II diabetes are significant for all aspects of our health. First, they are very likely to disrupt the gut–brain axis, exacerbating the addictive nature of food. Experimental models in mice suggest that diets high in fat and refined sugar cause inflammation in the small bowel and alter tryptophan metabolism, which increases growth factors and neurotransmitters in the part of the brain responsible for connecting senses and emotions to memory, called the hippocampus. This in turn changes exploratory and anxiety-like behaviours in mice.[16] However, inflammation will also influence every other major gut–organ axis in the body, with significant implications for our risk of cardiovascular disease.

In 1988, the US endocrinologist Gerald Reavan proposed that obesity where fat accumulates around our internal organs (known as central obesity), diabetes and high blood pressure were linked as part of the same disease, which he called 'Syndrome X' or the 'metabolic syndrome'. Almost one billion people across the world now suffer from this condition. Today we diagnose people with it if they possess three of the five following conditions: central obesity, hypertension, hyperglycaemia (high blood sugar), hypertriglyceridemia (bad fats in the blood) and low-serum high-density lipoprotein (protective lipids). The bigger question is: does the microbiome play a role in causing all these different components of Syndrome X through a common pathway?

To answer this, we are going to need to understand the important role played by the liver.

The inflamed liver – the gut–liver axis

The liver is a metabolic powerhouse. It's the only organ in the human body with greater metabolic potential than the gut microbiome, and it's been given multiple responsibilities for our survival. It processes haemoglobin, regulates blood clotting, controls amino-acid synthesis, detoxifies ammonia, metabolizes drugs, produces cholesterol and bile and converts glucose into glycogen for safe storage. Almost everything we ingest will make its way through the liver for processing at some point, so it plays a key role in regulating the composition of the gut microbiota, the integrity of the gut barrier and the health of our immune system. If the gut fails, so does the liver, and vice versa. Because of this, the metabolism of a junk-food meal by the microbiome plays an important part in the development of liver disease, which in turn defines our risk of obesity and diabetes. We are now starting to understand how specific components of junk food make this happen – and it's not what you think.

Because synthetic sugars like fructose inflame the gut, this damage means that bacteria and endotoxins pass into the liver, causing fat to accumulate (steatosis), inflammation and, ultimately, irreversible liver fibrosis, cirrhosis and failure.

Non-alcoholic fatty liver disease (NAFLD) is the most common chronic liver disease worldwide that results from this process, affecting 25 per cent of the global adult population and about 70 per cent of patients suffering from type II diabetes. Patients with NAFLD have a reduced gut microbial diversity from those without, and they also have a different faecal mycobiome composition from those with mild disease. Recent data suggest that patients with NAFLD have a very specific collection of fungi living in the gut, and that *Candida albicans* in particular plays an important role in regulating the gut's immune response that drives liver injury. When researchers colonized the guts of germ-free mice with the faeces of patients with NAFLD and then fed them a high-fat Western diet for twenty weeks, they induced fat deposits in the liver known as steatohepatitis. When these animals were then given antifungal medicines, their liver damage got better.[17] The very fact that fungi are able to grow like this in these patients suggests there is something fundamentally wrong in the gut ecology, because *C. albicans* has been able to outcompete its neighbours.

Unfortunately, we don't only abuse our livers with refined sugars and high-fat foods; we also drink, and 140,000 people die from alcohol excess each year in the US. Alcoholic fatty liver diseases have a lot in common with non-alcoholic ones – similar changes in the ecology and permeability of the gut, as well as shifts in the levels of bile acids, ethanol and choline metabolites and a local immune response in the liver. Some researchers have asked if the gut microbiome could be contributing to the development of non-alcoholic fatty liver disease through the production of its own alcohol. A team of Chinese scientists looked at the faeces of forty-three people with NAFLD alongside those of forty-eight healthy people,[18] and found significantly greater abundances of a strain of an alcohol-producing bacterium called *Klebsiella pneumoniae* in the patients with NAFLD compared to the healthy group. When these strains were given to mice as part of an FMT study, the mice also developed NAFLD. If these bacteria were eliminated from the FMT before it was given, the mice did not develop the disease. Our microbiome is literally drunk.

So how do we start to reverse these changes in the gut–liver axis that are causing us such harm through inflammation, liver injury and weight gain? Stopping drinking and going on a diet is the obvious strategy, and there are lots of diets to pick from. Choose wisely, because the impact of a radical diet on the microbiome may also have unintended consequences. High-protein, low-carbohydrate diets, such as the Atkins diet, or the more fashionable keto diet (high-fat, low-carb) have helped many individuals lose weight. They work by switching the body's dependence from glucose to water-soluble molecules produced from fatty acids in the liver called ketone bodies. This dramatically improves insulin resistance and leads to rapid weight loss. But just because these diets facilitate weight loss, that doesn't make them 'healthy'. Beyond the side-effects of 'keto flu', brain fog, diarrhoea or constipation, urinary frequency and vitamin deficiencies, these diets cause inflammation in the gut, because when animal fat or protein is the dominant source of energy for the microbiome, it generates cytotoxins and harmful metabolic by-products that are not good for the liver, and which increase the risk of chronic diseases such as cancer, as Stephen O'Keefe demonstrated in his study (see page 221).

A successful and sustainable weight-loss strategy needs fibre and balance, with some carbohydrate, and the goal should be to reduce inflammation in the gut. This is in part why intermittent fasting is so effective. However, the real trick to sustained weight loss through dieting is sustained behaviour change, which means that the gut–brain axis is a much more valuable target.

Maybe a better solution to surgery would be a faecal transplant? Early FMT trials in patients with metabolic syndrome have begun to demonstrate some improvements in insulin sensitivity. A group from Amsterdam was also able to derive a probiotic strain called *Anaerobutyricum soehngenii* from these studies, which appeared to correlate strongly with this outcome. When this single strain was used in a randomized control trial of twelve subjects with metabolic syndrome, glycaemic control was improved within just twenty-four hours.[19] This particular bacteria influenced an appetite-regulating hormone called GLP-1 in the duodenum that plays an important role in insulin secretion. This means that a whole FMT may not be necessary anyway, and that the type of bacteria in the donor sample is really important.

What patients eat after having received an FMT is also highly significant in defining its success. A recent study of seventy obese patients with metabolic syndrome randomized to an FMT with either a high- or low-fermentable fibre diet demonstrated that those on the low-fermentable diet (cellulose) significantly improved their insulin resistance.[20] This observation was independent of diet or medications. It may be that the cellulose improved the delivery method or promoted grafting, or even that it enabled beneficial microbes to bloom.

But is it all about the diet, or do other behaviours matter when it comes to optimizing the microbiome for weight loss or obesity prevention?

The sedentary microbiome

The amount of physical activity that the average person does has declined steadily with urbanization and digitization. A WHO analysis of 1.9 million participants from 168 countries found that more than a quarter of the world's population are not getting enough activity, which is defined as at least 150 minutes of moderate-intensity activity or seventy-five minutes of vigorous-intensity physical activity per week. So how does this contribute to our metabolic health, and does our activity influence our microbiome?

We spend most of our daily calorific intake simply maintaining our baseline metabolic rate, which is hungry work. The number of calories we actually 'burn' during exercise as a proportion of this work is depressingly small, although it depends on the type and duration of exercise that you perform. The good news is that exercise significantly improves almost all aspects of our health, and dramatically influences the function of the microbiome, which also benefits us in a huge number of ways.

First, regular exercise alters our microbiome's ability to metabolize our food. Mice that are allowed to run voluntarily on a wheel, for example, increase the diversity of the gut microbiome, the size of their caecum and the microbiome's metabolic potential, which produces more anti-inflammatory short-chain fatty acids in their stool, which is good for the gut and the brain.[21] And the more exercise, the better: when mice are made to exercise every day, they have a greater diversity than when they are left to do it voluntarily. This means that you should re-employ that personal trainer, because motivation may change your microbiome! The bad news is that if you don't care for your microbiome, it will influence your exercise potential. For example, antibiotics reduce voluntary exercise behaviours in 'athletic mice' that are specifically bred to run, which means that they exacerbate sedentary behaviours in obese individuals.[22]

In humans, the health benefits of exercise on the microbiome will vary, depending on what's in the engine at the start of exercise, the physiological status of the person exercising and their diet. Lean subjects are more likely to demonstrate changes in microbiome diversity than obese individuals, particularly over shorter periods of exercise lasting six weeks or so. This means it appears to be much harder for obese individuals starting exercise to get the microbiome into a fat-burning mode, and any changes will quickly revert when regular exercise ceases. All hope is not lost, however, and if you can keep going, then the microbiome will respond.

In a study of nineteen women undergoing six weeks of endurance training, researchers demonstrated distinct taxonomic shifts in the participants' microbiomes that were independent of age, weight, fat percentage, energy or fibre intake, and which included an increase in *Akkermansia muciniphila* and a reduction in Proteobacteria.[23] These observations were not influenced by menopausal status. Other studies in similar cohorts have identified that Prevotellaceae were positively correlated with peak power output and Christensenellaceae were negatively correlated with changes in weight.[24] Interestingly, probiotic strains of the Christensenellaceae family have recently been proposed to have anti-obesity potential in their own right.[25] Collectively, these observed changes in microbial taxa meant that it was possible to predict, with a good degree of accuracy, who was going to benefit from exercise versus who was not, based on a pre-intervention analysis of the microbiota. It also suggests that there may be some degree of gender dimorphism in the response of the microbiome to exercise, although we can say that exercise has benefits across all genders and ages, and we see similar observations in children.

For some members of society living at extremes of obesity or metabolic susceptibility, dieting and exercise are not enough and they remain at high risk of serious harm from diabetes or cardiovascular disease. In these patients, a more radical approach is required.

Surgery for the obese microbiome

Weight-loss surgery works through a combination of calorific restriction, altered gut-hormone function, altered bile-acid signalling and microbiome changes. It's more effective than current medication for the control of insulin resistance, and large studies from Scandinavia have demonstrated that morbidly obese people who have the surgery are significantly more likely to survive and that they will live longer than those who don't, although having the surgery still doesn't get people's survival rates to match those of lean individuals.[26] In theory, bariatric surgery is the perfect solution to a global crisis.

When you break a complex system there will always be unintended consequences of that destruction, no matter how carefully you go about it. Bariatric surgery doesn't only change the anatomy of the stomach, the small bowel and its microbiome – it impacts all of the body's microbial niches. In the gut the changes are most dramatic, and these happen in a similar pattern in both humans and animals. There are large shifts in the main gut phyla towards higher levels of Proteobacteria and lower levels of Firmicutes and

Bacteroidetes. The altered ecology of the microbiome and its co-metabolism of bile acids are important in explaining why this surgery works so well. So much so that it's possible to predict who will put weight back on after surgery, based on an analysis of the microbiome. And in animal models of bariatric surgery, antibiotics reduce the amount of weight loss after the operation.

These changes in the microbiome have important metabolic ramifications for the long-term health of the gut. For example, after surgery the microbiome begins to ferment proteins in much greater numbers, and some of the toxic by-products of this process damage the genes of the cells that line the gut. This may be one mechanism that explains findings from a Scandinavian bariatric database suggesting that weight-loss surgery increases the risk of colon cancer,[27] which is particularly surprising as obesity is a major risk factor for bowel cancer. Such surgery it also changes all the known and unknown critical metabolic and proteomic signalling cascades that surge down the gut's chemical superhighway, with significant implications for the gut–brain, gut–liver and gut–lung axes. Because our microbes metabolize our medicines, it also changes our susceptibility to their toxicity and side-effects. And there are other, perhaps more important implications.

Obese women have a lower prevalence of *Lactobacillus*-dominant vaginal microbiomes and a higher prevalence of a high-diversity vaginal microbiome, with increased levels of local inflammation. After bariatric surgery, those with the greatest weight loss at six months are most likely to have a *Lactobacillus*-dominant vaginal microbiome. The implication is that bariatric operations may inadvertently influence reproductive oncological and reproductive health for the better or worse.[28] More importantly still, should she choose to conceive, the changes in a woman's gut microbiome after malabsorptive weight-loss surgery persist throughout pregnancy. This may have a positive effect, such as reducing the level of inflammation or the risk of gestational diabetes. It also creates challenges, because it can contribute to reduced foetal growth or have as-yet-unknown signalling implications for the developing baby. A team from Imperial College London longitudinally studied forty-one mothers who had undergone weight-loss surgery and compared them to obese women having babies. When compared to obese women who had not had weight-loss surgery, they excreted significantly greater quantities of protein fermentation products in their urine, which is a sign that the gut microbiome is unhappy. The most striking observation was that if their mothers had undergone bariatric surgery, one of these microbial co-metabolites (Phenylacetylglutamine) was also statistically more likely to be excreted in the urine of newborn babies.[29] This is hugely important new information, which suggests that surgical weight-loss interventions can have transgenerational metabolic consequences that are mediated by the gut microbiome. The true consequences of this are not yet known, but the microbiome must now inform the debate about whom we perform this surgery upon, and at what age this becomes safe.

Bariatric surgery is largely a treatment for wealthy countries that can afford it, although it is the most deprived members of society who generally suffer from obesity. Perversely, the deprived are also much more likely to suffer from another harmful nutritional extreme: hunger.

Starvation and malnutrition

More than 2.3 billion people (or 30 per cent of the global population) lack year-round access to adequate food. There is significant global variation, with more than half of all undernourished people living in Asia and one-third living in Africa. For the 'have nots', the global famine deteriorated during COVID-19, and it is once more being inflamed by conflict. This is not only in Africa. In the UK, the Institute for Fiscal Studies found that food-bank users rose by almost 20 per cent between February and April–May 2020.[30]

In some respects, under-nutrition and over-nutrition are two sides of the same coin. They both result from ineffective food policy, deprivation, conflict, access to education and inequality. Their physical manifestations are the result of biological and microbial processes at different ends of the same spectrum, pushed to different extremes.

Malnutrition falls into three broad categories – namely underweight, stunting and wasting. Half of the malnutrition in the developing world is associated with diarrhoea or repeated infections of intestinal microbes, caused by drinking unsafe water and inadequate sanitation. Helminths are often the culprit, and these impair the physical and mental growth of children, thwart educational achievement and hinder economic development. Approximately 100 million children globally now suffer from stunting caused by malnutrition. Stunting of growth has its origins in nutrient deprivation during gestation and lactation, so the microbiome is heavily implicated.

Generally, children suffering from this condition have an increase in members of the Proteobacteria and lower levels of overall community diversity.[31] They're also more likely to have an increased prevalence of Enterobacteriaceae[32] and, collectively, these changes can be seen between six and twenty-three months prior to the development of growth stunting.[33] Phage isolated from stunted children are also distinct from those in healthy children, because certain phage evolve with Proteobacteria dominance in the developing gut. It's extremely likely, then, that dark matter plays an important part in shaping the bacterial community that influences the stunting phenotype. Most importantly, this opens an entirely new therapeutic window, and research is now ongoing that is seeking to engineer the microbiome to help treat malnutrition.

Kwashiorkor is the most extreme form of malnutrition, where protein deficiency often leaves starving children with swollen bellies that are full of fluid. Jeff Gordon's team (see page 236) studied Malawian twin pairs during the first three years of life, where half of the twin pairs remained well nourished, and the remaining half developed symptoms of malnutrition. Just like in the obese phenotype, genetically identical malnourished children respond to extremes of nutrition in very different ways, in part because they have different gut microbiomes.[34] When these were placed in germ-free mice by FMT, the symptoms of malnutrition could be replicated. It still blows my mind that malnutrition can be transferred between species simply by changing the gut bacteria.

Studies such as this one in Bangladeshi and Malawian children have demonstrated that the gut microbiome is slower to mature in malnourished states. The problem, as Gordon's team discovered, is that children with severe acute malnutrition may never get back to a healthy gut, even when the correct nutrition is provided. Amazingly, co-housing growth-stunted mice with mice that had received microbes from non-stunted humans reversed the stunting effect.[35] So does the whole gut-microbiome community need to be present in the first place, for stunting to happen?

The team observed that a critical functional unit of bacteria – which they referred to as an 'eco group' – may in fact be all that is needed. This was a discrete group of bacteria of only fifteen taxa and it was stable over time in both animals and humans.[36] This was important, because if only a small group of bacteria is having the majority of the nutritional benefit, an FMT is not needed to get the microbiome back on track. It should be possible to give just these fifteen bacterial species. This is exactly what the scientists did.

They developed what they called a microbiota-directed complementary food (MDCF), specifically designed to meet the requirements of the microbiome in malnourished children. A randomized control trial was performed that compared the MDCF to a meal of equivalent calories in 123 slum-dwelling Bangladeshi children, with moderate acute malnutrition, between the ages of twelve and eighteen months. The results showed a moderate but significant improvement in the children's weight-for-length and weight-for-age scores when compared to the control diet.[37] Some people were critical, as the sugar content of the MDCF was high, but this most certainly represents a hugely important step because it means we can develop low-cost microbiome-targeted nutritional strategies that have nutritional benefit and which don't require the complicated and expensive paraphernalia of FMT.

One word of caution, however. In our rush to help, we must not cause further unintentional problems. There is now a double burden of malnutrition prevalent in countries undergoing 'nutrition transformation' where under-nutrition and obesity coexist, often because of the Westernization of the diet that Burkitt feared. We're in danger of burdening low- and middle-income countries with more Western disease, which they can most definitely do without.

GEM disease

The depressing fact is that obesity and malnutrition are not extreme phenotypes or even rare events. They are the norm. The dramatic changes to our exposome and our microbiome have overwhelmed our genome, and we have not been able to adjust – and the result has been an inflamed gut. This is not a 'metabolic syndrome' at all, but an inflammatory syndrome; obesity is simply one disease of many that stems from this common pathway, and which has so dramatically changed our risk of all chronic disease.

The problem is that there is not one single causative factor that we can simply reverse to solve this problem, and the exposome has changed our metabolism over generations. The meta hypothesis therefore helps us make some sense of how the obesity pandemic has occurred over the last fifty years, and why it is linked to so many other health crises. There have been four key events:

1. *The maternal microbiome has been disrupted by antibiotics, medicines and a Westernized diet over successive generations.*
   This means that the orchestra can no longer remotely programme the developing immune system in an optimum fashion, creating susceptibility to obesity in our offspring.
2. *After birth, these same forces have compounded the incomplete assembly of our children's microbiomes and have exacerbated this problem, which is then much harder to correct.*
   During childhood, the resulting inflammation in the gut modifies our behaviours and dulls the effectiveness of the hormones that regulate our blood sugar and fat metabolism, further priming us for diabesity. In the

twenty-first century the processed Western diets that we feed to our children in their early lives imprison them in a cycle of malnutrition, obesity, metabolic disease and insulin resistance that they cannot escape from.

3. *In adulthood the gut microbiome adapts to a Western diet, pollution and xenobiotics, optimizing its ability to harvest energy.*

   Because our symbionts are either completely missing or impotent, they're no longer able to compete, so the gut blooms with amensalists and even pathogens. This explains the rising risk of chronic diseases such as hypertension, hyperlipidaemia, NAFLD, cardiovascular disease, mental-health problems and cancer.

4. *In old age, our bloated microbiome lacks the resilience to manage the exposome, and protein plaques form in our nerves, misting our memory of all this.*

In just a few generations, the only option for those trapped by this damaging cycle has been bariatric surgery. The problem is that this is not scalable or sustainable, and it creates its own set of biological complications that aren't well understood. Prevention must be the strategy.

Maybe what we really need is an obesity-targeted, microbiota-directed complementary food. Or perhaps it's something else, something more profound. Our guts are an ecosystem worthy of protection – exactly like a rainforest or any other precious biodiverse resource at risk from climate change. We can only improve things if we take collective responsibility. This is not impossible, but it is hard. We can create evidence-based, effective dietary and antibiotic prevention strategies that focus on optimizing maternal gut health and the developing gut, and on reducing inflammation in the adult gut. This is worth doing, and I hope that one day bariatric operating rooms will become a historical artefact.

15. Hacking the Symbiont

Having a deep insight into how the microbiome causes disease is helpful. However, this needs to be turned into actionable information that we can all leverage to meaningfully improve our health. We do have several therapeutic options at our disposal, such as taking a probiotic, prebiotic, postbiotic, synbiotic or next-generation probiotic (see below). Exercise can also be considered as microbiome therapy. The right approach for you will depend on your health needs and objectives; the next decade will be spent finding new ways to target these therapies in a more personalized way.

For most of us, though, the simplest way is to change our diet, which usually means consuming specific foods or removing others. This is particularly challenging when we are unwell, not least because the dietary advice given to patients by doctors is often conflicting at best, or plain wrong at worst. And this has serious implications for our health, because the microbiome changes drug efficacy and toxicity, and diet has therefore confounded almost every drug trial that has ever taken place in a human. Most diet–drug–bug interactions are poorly understood because of the tremendous complexity of the gut microbiome and the vast catalogue of enzymatic functions that it has at its disposal for altering the chemistry of a particular pharmaceutical compound. These vary with the nature of the diet and with the disease being treated.

Food is medicine

Some diet–microbiome–drug interactions are precisely defined and are specific to certain strains or pharmacological agents. For example, the drug digoxin is derived from the foxglove plant, and is used in the treatment of cardiac failure and cardiac arrhythmias. In the 1970s doctors started to notice that the effectiveness of the drug varied significantly between urban and rural dwellers, and in patients at extremes of life or those on antibiotics such as erythromycin. In humans, this happens because of a strain of bacteria called *Eggerthella lenta*, which possess a gene coding for a CYP450 enzyme that metabolizes the drug. However, this particular enzyme is also inhibited by an amino acid called arginine. Studies in germ-free mice colonized with *E. lenta* showed that a diet high in protein only altered the metabolism of digoxin in animals given the *E. lenta* strain that possessed this gene.[1] This means that we should really be offering dietary advice to patients taking one of our most commonly prescribed cardiac drugs.

This example is a rarity, though, and most people who are unwell are not lucky enough to have this insight. In fact many patients undergoing hospital treatment rely on their nutritional supplements or intuition to support conventional medical therapy. We still have limited mechanistic evidence to describe how these supplements interact with pharmacological agents, or how they should be dosed and in whom. As a result, modern Western medicine largely rejects them, leaving them on the spice rack for alternative practitioners. Spices contain polyphenols and other bioactive molecules, and they most definitely modify our microbiome. Yet only a small number of spice–microbiome interactions – for instance, those for pepper, ginger, garlic, cinnamon, saffron, nutmeg and turmeric – have been defined.

Turmeric is rich in bioactive molecules with anti-carcinogenic and proposed anti-inflammatory properties. In mice models of colitis (inflammation of the bowel), it increases the abundance of probiotic bacterial strains associated with tryptophan metabolism, such as *Lactobacillus* and *Clostridia-UCG-014*, which strengthen the barrier function of the gut and reduce the severity of the inflammation.

In a similar mouse model, cinnamon oil improved gut microbial diversity and the richness of intestinal microbiota, and increased the abundances of short-chain fatty-acid-producing bacteria. In mice fed a high-fat diet, whole-garlic supplementation improved the bacterial diversity of the gut microbiome, especially increasing the relative abundance of Lachnospiraceae and reducing the relative abundance of *Prevotella* as well as the level of circulating cholesterols. Spices do significantly change the metabolism of the gut and it's entirely plausible to me that they have a health benefit, if they are taken in a specific manner for a specific purpose – but not when they are plonked on top of a cheeseburger.

Many patients use supplements to support their traditional cancer chemotherapies. Recently, researchers from Macau University of Science and Technology discovered that the polysaccharides found in ginseng have a very modern use in the treatment of cancer. Specifically, they identified that it improved the effectiveness of PD-1 checkpoint inhibitors in mice. The reason this occurs is because the specific polysaccharides found in ginseng are co-metabolized by gut bacteria.[2] This produces valeric acid and reduces the production of kynurenine, which in turn changes the expression of T cells in the tumour microenvironment. The net result is that the tumour is sensitized to the drug.

When it comes to cancer, what we don't eat can be as important as what we do eat. Restricting essential nutrients such as glucose, glutamine, glutamate, asparagine, aspartate, methionine, serine and folate has been found to limit cancer growth, even though the nutrients work through different mechanisms. The workings of the microbiome might also explain why intermittent starvation influences traditional drug treatments. For example, fasting combined with chemotherapy in the treatment of breast-cancer patients reduces the toxicity of the treatment,[3] and it reshapes systemic and intratumour immunity and activates antitumour immune responses.[4] Metformin is a widely used drug for the treatment of type II diabetes, and there is emerging evidence that this drug reduces the incidence of breast and pancreatic cancer and improves survival from colorectal cancer. While it deprives the cancer of the sugar it needs to grow, metformin also interferes with the cancer's ability to make proteins (through the mTOR pathway). Its effect is enhanced when the patient restricts their food intake because, collectively, these approaches also restructure the microbiome:

metformin alters the composition of gut microbiota through the increase in mucin-degrading bacteria such as *Akkermansia muciniphila*, as well as several short-chain fatty-acid-producing microbiota.

Intriguingly, specific dietary strategies such as the keto diet may have a role to play in cancer therapy. This is because, in an evolutionary quirk of biology, cancer depends on sugar as its primary source of energy – a ketone diet will not only deprive it of this, but will also reduce insulin resistance. It also increases inflammation in the gut, which under normal circumstances would be a bad thing. However, this could potentially be a good thing if you're trying to make a tumour microenvironment more sensitive to drugs like immunotherapy. The water-soluble molecules produced from fatty acids in the liver during a keto diet are called ketone bodies. One of these, called 3-hydroxybutyrate (3HB), slows tumour growth by upregulating T-cell functions in the tumour microenvironment where checkpoint inhibitors work. Amazingly, a keto diet was able to rescue mice fed on a standard diet that had failed to respond to a type of immunotherapy known as a checkpoint inhibitor (called anti-PD-1). Scientists are now starting to identify particular species such as *Eisenbergiella massiliensis* that appear to be responsible for the rise in blood concentrations of 3HB, suggesting another microbiome target for cancer therapy.[5]

Intermittent starvation is also helpful in reducing inflammation in the gut, and this can have therapeutic benefit. For example, the neurotransmitter serotonin, which is made by bacteria, blocks the process of autophagy (the recycling of dead or dying cells) triggered by starvation in the gut. This in turn increases the gut's susceptibility to inflammation, known as colitis. If autophagy is stopped by removing the gene that controls it, and the dietary supply of serotonin is limited, then the microbiome becomes pro-inflammatory and this causes colitis, because neither the serotonin nor the autophagy can keep it in check.[6] Patients with colitis have been found to have low levels of serotonin in the blood, and when this was replaced, their colitis improved. In short, if you take the brake off autophagy, the gut microbiome will run riot, and vice versa. The microbiome can therefore be directly manipulated to reduce inflammation in the gut.

This sort of dietary strategy isn't always easy, and a better option for some is to selectively remove components of a diet that bacteria use to create harm. For instance, numerous studies show many patients report that specific foods exacerbate the symptoms of IBS. One of the most popular diets to treat this is the restriction of short-chain fermentable carbohydrates, known as the low fermentable oligosaccharide, disaccharide, monosaccharide and polyol diet, or FODMAP. This restricts multiple fermentable oligosaccharides (fructans, galacto-oligosaccharides), disaccharides (lactose), mono-saccharides (fructose when in excess of glucose) and polyols (for example, sorbitol and mannitol). Excluding these fibres reduces the amount of gas produced when these fibres are broken down by bacteria, and eases symptoms of bloating. The FODMAP diet might also work because it changes the microbiome and the metabolites it produces. For example, the diet typically leads to a reduction in the relative abundance of *Bifidobacteria*, *Faecalibacterium prausnitzii* and *Clostridium* cluster IV and there's even the suggestion that bacteria may predict treatment response.[7] These bacteria reduce the sensitivity of the bowel to the sensation of stretching as it bloats, probably through the production of neurotransmitters via the enteric nervous system.

However, many of these dietary solutions are used as part of a multidisciplinary approach. As a result, patients coming to see me in my clinic will also often ask the question 'Should I take a probiotic?'

Probiotics

At the time of writing, more than 38,000 papers have been written on probiotics, and they have been trialled in every conceivable clinical condition. There's strong evidence for their use in preventing traveller's diarrhoea, antibiotic-associated diarrhoea and for the treatment of gastroenteritis in children. There is also good evidence for their use in the treatment of constipation, bacterial vaginosis and in patients having surgical treatments on the gut, but we lack a guide for specific probiotic strains that can be used under specific clinical conditions. Their clinical efficacy is also variable because our gut ecology varies with age, diseases, medicines or changing environments.

A probiotic is officially defined as 'live microorganisms that, when administered in adequate amounts, confer a health benefit on the host'.[8] In other words, the bacteria must be alive when they reach the part of the gut they'll be working in; they have to be there in sufficient numbers to do the job; and the mechanism by which they improve our health must be known. Proving that you've met all three criteria is a very hard task, because the human gut can't be readily accessed, and the probiotic is working in a crowded ecosystem. Because of this, many probiotic manufacturers aren't allowed to make numerous health claims about their products.

In some countries, such as the US, medicinal probiotics are often referred to as Live Biotherapeutic Products (LBPs) by regulators, in a bid to describe their roles when it comes to treating diseases more accurately, and to differentiate them from consumer food products. So we're talking about live organisms, such as bacteria, that we can use to prevent, treat or cure a disease or condition . . . but they're *not* a vaccine. It's confusing!

The most used probiotic genera include *Lactobacillus*, *Bifidobacterium*, *Streptococcus*, *Enterococcus*, *Escherichia*, *Bacillus* and the yeast *Saccharomyces*. However, there are more than 260 different phylogenetically discrete species from the *Lactobacillus* genus alone. Although they may share some common functions, each probiotic species is actually a discrete entity that will interact differently with our gut. So the health properties of *Lactobacillus casei* may be subtly different from those of *Lactobacillus reuteri*. Companies manufacturing these brands will go to great lengths to patent specific strains of each, then rebrand them with commercial names so that you may never know their true origin. There are now somewhere between 500 and 1,000 strains of probiotics on the market.

Commercial preparations should quantify the product in

colony-forming units (CFU), which describes the total number of viable cells in the product. Generally, the more CFU you can get, the better. I would typically recommend using a product with at least 1×10^9 (1 billion CFUs) or 1×10^{10} (ten billion CFUs), although some can contain up to fifty billion CFU or more. If you're unsure what specific strain will be best for you, I'd recommend using a multi-strain preparation.

Probiotics don't reset the microbiome, nor do they lead to wholesale changes in the ecology of the gut. In fact their impact is subtle. There may be one billion bacteria in your dose, but there are 100 trillion in the gut. So they need to be used as part of a dietary strategy that's trying to fix a specific problem.

Globally the regulations that define manufacturing quality and dose vary, and so do dosing quantities and strategies. Breakthroughs are at long last coming in the selective targeting of these live biotherapeutic strategies, but this field is still overwhelming for consumers who simply cannot be sure that the strains they need are the ones they are getting, or that they have a proven clinical benefit through high-quality randomized control trials in humans.

We now need national formularies that provide this information for both clinicians (most of whom are equally bemused) and consumers, so that we can start to make informed choices based on mechanisms, safety and potential benefit. It is time that the industries that profit so handsomely from these products help to support these initiatives, and that our governments lead them. Without this, we cannot have a ground truth for pushing the field forward. Once you've chosen your brand, you need to commit to it, to allow the bacteria to engraft or colonize the gut. This means for at least four to eight weeks, and if it works, you must keep going. If it doesn't, save your money and move along.

Technically, fermented food can't officially contain 'probiotics' unless the manufacturers are able to make confirmed mechanistic health claims about specific strains or combinations of strains. However, I find them very useful, and I think they are worthy of a brief review.

Kefir and fermented dairy products

In Europe, some manufacturers can make health claims about live yoghurt cultures, based on the presence of a lactase enzyme found in combinations of bacteria such as *Lactobacillus delbrueckii* subsp. *bulgaricus* and *Streptococcus thermophilus*. This is because studies have shown that fermented dairy products reduce the risk of type II diabetes, weight gain and mortality, and other studies suggest that they prevent high cholesterol (hyperlipidaemia).[9] But the quality of the evidence we have varies, depending on the type of fermented food, and it's not always clear which bacteria or part of the 'food matrix' is actually benefiting us. This is compounded by variation in the ingredients used and the fermentation process (temperature, pH, oxygen concentration and duration of storage).

Fermented milk from the Caucasus is called kefir, and it is now big business; the global market is predicted to be worth US $1.84 billion by 2027.[10] Where the milk comes from will make a big difference to the final strains present in a given batch and its precise health benefit – though we're starting to see some guidance on minimum standards. For example, the Food and Agriculture Organization (FAO) and WHO state that kefir grains should contain a minimum of 10^7 colony-forming units (CFU)/g microorganisms and the final product should contain at least 10^4 CFU/g of yeast. Most strains found within kefir are *Lactobacilli* (LAB), a genus that occurs with generally low abundance in the human gut, its variable presence being affected by age and lifestyle. It's hard to work out exactly which of these strains is having the benefit, or if it's a network effect. Nor do we fully understand how they interact with the human gut microbiome and whether they can all survive the physical and chemical challenges of the gut.

A different strategy to drinking fermented milk once a day (which I highly encourage) is to consider a targeted probiotic that's been engineered for a specific task.

Next-generation probiotics

The massive growth in metagenomic sequencing, omics technologies and genetic engineering means that we're about to push probiotic technology into a whole new dimension. First, many of the strains that I have described in this book, which were discovered through the application of these technologies to disease states, will gain regulatory approval to be used as therapies with defined health benefits. For example, *Akkermansia muciniphila*, *Faecalibacterium prausnitzii* and *Anaerobutyricum hallii*.

But not only are we discovering many more strains, we're also discovering their functions and how to manipulate them for our own benefit. The age of synthetic probiotic modification has arrived. And unlike current probiotic strains, these products – known as 'next-generation probiotics' (NGPs) – aren't considered foods, and will be delivered under a drug regulatory framework.

We're already seeing significant advances in our ability to manipulate microbial genomes for our benefit, and NGPs are being used either as a monotherapy or in combination with other anti-cancer therapies. *Escherichia coli* Nissle 1917, a probiotic that commonly colonizes cancers, has been engineered to also produce a metabolite called L-arginine. When this enters a cancer cell, it produces ammonia that activates the immune response, drawing in T cells, killing the cancer. This works in parallel with immunotherapy drugs to increase their anti-cancer effects.[11]

Probiotic strains can also be used as vehicles to transport anti-inflammatory molecules right to the part of the body where they're needed. *Lactococcus lactis*, commonly used in cheese production, can be genetically altered to make an anti-inflammatory enzyme called a serine protease inhibitor, which treats inflammation in an animal model of colitis. The common colonic bacterium *Bacteroides ovatus* has also been engineered to produce an anti-inflammatory molecule called TGF-β1, which was as effective as steroids in animal models of colitis.[12] We can even engineer bacteria to address some of the biggest challenges in the twenty-first century by programming them to clean up our internal organs and remove toxins, plastics, xenometabolites and other harmful molecules. Live biotherapeutics, for example, can safely degrade antibiotics in the gut to prevent their associated impact on the intestinal diversity.[13] *E. coli* Nissle 1917 has been engineered to demonstrate prophylactic and therapeutic activity against *Pseudomonas aeruginosa* during gut infection in two animal models.[14] And as we've seen, probiotics are also being targeted at specific organ functions – such as psychobiotics for brain health.

And this revolution will not stop at probiotics.

Prebiotics and postbiotics

A prebiotic is, usually, a fibre that serves as a nutrient for beneficial microorganisms living within the gut that in turn have a direct health benefit. While many prebiotics are fibres, not all fibres are prebiotics. In our diet, prebiotics are more commonly found as insoluble fibres like inulin (which you can find in asparagus, bananas, burdock, chicory, leeks, garlic and Jerusalem artichokes), fructo-oligosaccharides (FOS) and galacto-oligosaccharides (GOS), resistant starch, polydextrose, xylo-oligosaccharide (XOS) and isomalto-oligosaccharide (IMO). Other substances, such as polyphenols and polyunsaturated fatty acids (PUFAs), also function as prebiotics.

Like probiotics, prebiotics come in a diverse set of formulations, and you can even put them on your wholegrain high-fibre cereal. They can also be administered directly to other microbially colonized body sites, such as the vaginal tract and skin. Like probiotics, they have many benefits to the gastrointestinal tract (for example, inhibiting pathogen growth and stimulating the immune system), they benefit cardiometabolism (by reducing blood lipids and improving insulin resistance), mental health (for instance, metabolites that influence brain function, energy and cognition) and bone health. Some people, particularly those with IBS, require a detailed strategy for their fibre selection, as prebiotics can cause bloating.

You can take a prebiotic and a probiotic together, so that the former provides the strains that the latter needs to sustain itself in the gut. We call this combination a 'synbiotic'. Another approach is to try a 'postbiotic', or a food containing either inanimate microorganisms or their components, which confers a health benefit on the host.[15] For example, why grow the bacteria that metabolize fibre into short-chain fatty acids when you could simply give the short-chain fatty acid that has the benefit? The answer may be that it is more about where in the gut it is produced, the amount that is produced or the secondary benefit of the network effect of live bacteria. Studies examining the benefit of heat-killed bacteria for gastrointestinal, metabolic and cardiovascular conditions are now beginning to emerge, and it is likely that we will see more of these products on the shelves of health-food shops in the near future. Whether they actually work or not remains to be seen.

The side-effects of pre-, pro-, post- and synbiotics are usually minor and consist of self-limited gastrointestinal symptoms, such as gas. In a few cases, in individuals who were severely ill or in whom the immune system wasn't working, the use of probiotics has been linked to bacteraemia (bacteria in the blood), fungaemia (fungi in the blood) or infections that result in severe illness.

However, it's not only eating and supplementing your diet with synbiotics that can help you keep your microbiome in check. There is another very important way to ensure it stays healthy.

How to train your microbiome

One of the most important things you can do for your gut and your gut–brain axis is exercise regularly. Exercise, when taken with a balanced diet, can also be used to hack the microbiome.

The gut microbiomes of professional athletes contain greater abundances of Veillonellaceae, *Bacteroides*, *Prevotella*, *Faecalibacterium*, *Methanobrevibacter* or *Akkermansia* than non-athletes like me, which may have competitive benefits and explain why I have yet to win the Tour de France. The precise advantages these bacteria provide to elite athletes are now being defined, and there is evidence that regular exercise not only improves the health of the gut, but may also improve performance.

For example, researchers have identified that the bacterium *Veillonella atypica* was enriched in the guts of runners taking part in the 2015 Boston Marathon.[16] The researchers hypothesized that this may have helped improve the endurance of the runners. After long periods of exercise, a metabolite of anaerobic work called lactate starts to build up in the muscle. This is what causes the pain when you are training! They compared the function of *V. atypica* to another probiotic strain called *Lactobacillus bulgaricus*, which does not metabolize lactate. Fascinatingly, the mice given the *V. atypica* demonstrated an increased exhaustive treadmill run time, which means they could run further. When the short-chain fatty-acid propionate was injected into the rectum it had a similar effect, leading the authors to suggest that *V. atypica* improves physical performance by converting exercise-induced lactate into propionate, thereby identifying a natural, microbiome-encoded enzymatic process that enhances athletic performance.

Given the global variation in the microbiome, *where* the marathon is run is also important to any analysis. Runners taking part in the 2016 Chongqing International Half Marathon demonstrated increased species richness after running, with raised abundances of Coriobacteriaceae and Succinivibrionaceae.[17] As with changes in the global microbiome, each country or region may need to adjust to its own particular microbiome, and each event may require a subtly different optimization of the gut to achieve peak performance.

Good sh*t

A healthy microbiome is a diverse microbiome that has plasticity, resilience and the complete set of core metabolic and immune functions that we need for growth and the maintenance of health. There are simple things we can all do to improve our microbiome's biodiversity and optimize its function across the duration of our lives.

The best way to do this is by changing our exposome, and here are three points that can help.

1. *Nurture: help the microbiome assemble and give it the best start in life that you can.*

    If you are having a baby, think about what you want the microbial orchestra to play. Breastfeed if you can, but don't feel guilty if you can't. Deliver your baby however is safest. The microbiome is dependent on social interaction and play, so put your screens down. Travel, get outside and connect with the world around you. Get your hands responsibly dirty, and play with animals. Kiss and hold hands. Educate your children on the importance of looking after the precious ecosystem within us. Bring plants and life into your home and nurture them too.
2. *Feed: de-Westernize your diet and your lifestyle.*

    Move to a plant-based diet. Eat at least 30g of fibre a day (or thirty portions of vegetables per week), delete your fast-food delivery app, cook your own whole food and share it. Eat colourful foods with polyphenols and spices, and change up your menu from time to time. If you do eat meat, cut out animal fats, eat less red meat (no more than one portion a week) and preferably don't eat any processed meat. Stop eating refined sugar, but balance your carbs, as you do need some. Variety is key. Add fermented foods to your diet and eat them regularly.
3. *Protect: avoid taking antibiotics or medicines that you don't need, in particular painkillers or drugs that alter the pH or acidity of the gut (but ensure you consult your doctor before making any changes to prescribed medication).*

    Stop smoking and no vaping. If your BMI is too high, ask for help. Fad diets are generally bad for the gut microbiome, so don't bank on them. Exercise regularly in whatever way works best for your body and your needs, and ideally do some exercise that gets your heart rate up at least three times a week. Sleep is important, so remove technology from your bedroom. If at all possible, avoid places with high levels of air pollution. Consider taking a probiotic if you have a specific health need that would benefit. Manage your stress: the gut–brain axis sets inflammatory tone – you may need help to do this. Wear sunscreen, just as Baz Luhrmann said.

If this all sounds like common sense, that's because it is. However, I want to be clear that exercising and occasionally taking a probiotic or prebiotic cannot fix the changes to the microbiome in the face of overwhelming pressure from the modern exposome. Our microbiome can only respond to the environment that we choose for it, and there is no magic bullet. This framework must be adjusted depending on where you are on your GEM line, and on the specific needs of your gut microbiome at that moment in time. In the future, however, we may have even more tools to help us protect the microbiome.

Engineering the microbiome

It won't be long before an FMT becomes a commercial product that can be prescribed and collected at your pharmacy. As we begin to refine how FMT works, it may be that we'll no longer need to transport an entire ecosystem between individuals – we'll be able to transplant only the specific microbial functions that we need, and we will be buying microbiota-directed complementary foods. We will be able to choose the type of donors we want or the strains of microbes we need, which are tailored to our gut, based on a genomic or functional analysis. There will be a refined offering of dose, type and route of administration, like taking a pill. In this vision of the future, FMT is going to become a precision therapy. It's also possible we'll be storing our own microbiomes in times of health, to be used as treatments during times of disease. And the revolution that will enable this vision to occur is happening right now.

16. The Trillion-Dollar Product

An explosion of start-ups are now testing, mining, biobanking and engineering the microbiome in every field, from climate change to food production and human health. By the end of 2021, sixty-four microbiome start-ups had raised a total of US $1.6 billion[1] – and they aren't just coming for your gut. They want your brain, lungs and skin, and the microbiomes of every niche on and in your body. These companies don't simply want to make you healthier; they want to make you even more attractive. The cosmetic pharmaceutical (or 'cosmeceutical') market will engineer your microbiome to make your skin younger, your hair shinier and your body odour more alluring. We'll be able to discretely switch on genes in microbes that are already in your gut, lungs or genitals or on your skin. For this to happen, another revolution is needed.

The digital microbiome

Future doctors will need to practise a form of data-driven, emotionally intelligent precision medicine that places as much emphasis on food and the exposome as it does on pharmacology or surgery. To do this, we need a way of measuring microbiome health, and a new standard model of medicine that predicts how structural and functional components of the microbiome behave across all four stages of the holobiont's life. The microbiome has to be digitized.

Microbiome research is a world of clouds, databases, sequencing, germ-free animals and, occasionally, a Petri dish. It owes its rapid evolution to advances in computing power, but this is merely the beginning. We're moving towards a world of integrated systems biology, where the microbiome will be just one more omics science in the great algorithm that will explain the workings of all life on

Earth. But there's something particularly special about the microbiome that means it will be mined in its own right.

The microbiome is a rich source of new drug candidates that can resupply the dwindling pharma pipeline. We even have a catchy new word to describe this science. 'Pharmacomicrobiomics' is the future of personalized medicine, bringing together GEM interactions that define the efficacy and safety of our future medicines. These drugs won't be targeted at human genes – they'll be designed to work on particular bacteria or functions of the microbiome, for human benefit. Scientists across the world are already engineering bacteria that manufacture drugs, enzymes, hormones, antibiotics, toxins and metabolites. These will be able to kill or inactivate bad bugs or cancer, or even instruct other bacteria and yeasts to start producing molecules that treat or prevent chronic human disease. These small molecules will be able to disseminate to any organ in the body, including the brain. It's very likely that future treatments for neurological conditions will leverage the gut microbiome, and it gives us hope for conditions where we currently have no effective treatments.

Massive searches for the functions of our gut bacteria, archaea, fungi and viruses are now under way. We're attaching bar codes to spores to track them across environments, and we're generating artificial-intelligence methodologies that will give us deeper insight into how microbes will behave, both in and outside our bodies, and how they will interact with our diet, our drugs and the rest of our exposome. Large-scale *in vitro* bioreactors already let us simulate the entire gastrointestinal tract, but engineered 'organ-on-a-chip' models will allow us to test and develop drugs in the laboratory for humans that account for the microbiome and finally do away with slow, expensive and ethically challenging animal experiments.

Bacteria are also being digitized in a more literal sense. They are being seamlessly woven into bioelectronic devices that can be absorbed or embedded in the body to monitor our health. Probiotic biosensors can detect the presence of gastrointestinal bleeding and send the information wirelessly to a clinician.[2] The DNA of *E. coli* has even been engineered so that colonies of this bacterium work as a computer, making genetic calculations that solve real-world problems, like how to navigate through a city. Biofilms from the mould *Physarum polycephalum* are also being used to make computer chips that could replace silicon. Our future computers may be living systems made from microbes.

If we are to unlock this amazing potential, we need to have access to both genetic and environmental data over the course of human lifetimes and across populations. To capture this, the microbiome is being integrated into our digital lives through mobile apps that seamlessly record our health information, while companies use data from our home-testing kits to sequence our faeces. The microbiome has become a commercial commodity in its own right, and many microbiome start-ups hold personal and valuable data about every aspect of our health. Some of these companies are banking on the fact that the true value of this is not yet known. Not all of them are transparent about what's going to be done with the data, and they will not be accountable to us or anyone else.

If we're not careful, the future could be a more pathogenic strain of what the American philosopher Shoshana Zuboff calls 'surveillance capitalism' – an economic system centred around the capture and commodification of personal data for the purpose of profit-making. The microbiome-research world must ensure that its science doesn't end up controlled by the few, accessible only to those who can afford it. It must be open to those who need it most.

The opportunities for growth are undeniably huge. Through the microbiome we can once more compensate for our bad choices and begin to address the epidemic of GEM diseases. For this, however, we may need to take one further step into the unknown.

Synthetic biology

In 2010, Craig Venter's team (see page 11) claimed to have created the first form of synthetic life. They designed, synthesized and assembled a bacterial genome for *Mycobacterium mycoides*,[3] and then a synthetic genome led to the construction of a minimal bacterial cell and recoded *E. coli*,[4] and finally a yeast genome with the ultimate goal of being able to synthesize any organism. The vision is that we can design a microbe or modify its functions using a simple piece of software code for creating microbial genes.

For now, it's not possible to boot up a complete bacterial genome for any species that is not a genus of bacteria called mycoplasma, and you may be relieved to hear that it is most definitely not possible to synthesize chromosomes of higher eukaryotes. It is too expensive and there are major technical roadblocks in the way. It would require, for example, the high throughput synthesis of trillions of oligonucleotides, the building blocks of DNA. However, this is very likely to be solved with time through the development of scalable semiconductor chips, which means that the future of microbiome science may well be owned by global computing and data-science giants like Microsoft, Amazon and Google.

The synthetic genomics revolution that enabled this work started in viruses, and it gave birth to the mRNA COVID-19 vaccines that transformed our response to a global catastrophe. The potential of synthetic genomics is, though, much greater than simply targeting pathogens that cause pandemics. For instance, we're rediscovering old biotherapeutic technologies that were once discarded for reasons of biopolitics, like phage. This technology has lain relatively dormant in the West for a hundred years. Phage are not only a possible solution to the growing antibiotic resistance crisis, they also lend themselves to bioengineering human and planetary microbial systems. And we're already finding more immediate uses for phage in the treatment of cancer. They can be bound to chemotherapy drugs, for example, to target bacteria growing in cancers so that these tumours become more accessible to chemotherapy.[5]

Future synthetic microbes will clothe us and create the biofuel that will service our daily living. A mycelium's fast-fibres can already be grown quickly to produce materials used for clothing, synthetic leather, plastic alternatives and even synthetic organs. Fungi and yeasts have been used for thousands of years in our food production and will only become more popular, because we're not making enough food to sustain the needs of a growing population. Global demand for macroalgal and microalgal foods is also 'blooming', and algae are increasingly being consumed for functional benefits beyond the traditional considerations of nutrition and health. They're being added to synthetic meats to provide not only protein, but also texture . . . It's algae that make synthetic meat 'bleed'.

Could we leverage this technology to give our planetary microbiome its own environmental FMT? It's not impossible. Next-generation probiotics will also be made for aquaculture, coral and plant systems and the agribiome. Biofertilizers that promote the diversity and richness of the soil microbiome, and those that optimize the gut health of our livestock, can herald a new dawn of microbiome stewardship. This is critical if we are to solve humankind's biggest self-imposed problem – namely, climate change. Synthetic microbes will metabolize our plastics, recycle our carbon, nitrogen and sulphur and regenerate our environment. They will even help take human explorers to new planets as we seek fresh habitable environments.

Mary Shelley, the author of *Frankenstein* (the first great AI novel), would be nervous watching all of this unfold – although I'm also pretty sure she would be vaccinated. This brave new world brings with it major ethical, moral and biosecurity challenges that will need to be faced and overcome by regulators, scientists, industry and the public. Despite all the supposedly failsafe mechanisms for what is known as biocontainment that geneticists can code into their new life forms, such as self-destruct or inactivation functions, evolutionary biology tells us that the microbiota will find a way to navigate past our barriers. Carl Woese, whose RNA hypothesis gave birth to phylogenomics, was a religious man and by all accounts did not appreciate humans assuming they could explain the complexity of God's work.[6] I'm not sure what he would have made of Stéphane Bancel, the CEO of Moderna, claiming that its mRNA technology had unlocked the 'operating system' for life. For me, this is a naïve statement, because it forgets one very important thing: *Pratītyasamutpāda*. All things arise in dependence of other things.

Epilogue: How to Save the Microbiome

The human microbiome allows us to rephrase our understanding of why we suffer from chronic diseases. Raymond's scarred microbiome (see page xvi) had contributed significantly to his many medical problems, yet it was ignored until he was finally taken hostage by an amensalist. While he was lucky enough to respond well to a faecal microbiota transplant, an FMT is not a panacea and should be the last resort for the treatment of any chronic disease.

If you're taking a medicine, then the foods you eat, the exercise you take, the sleep you get and the weight you carry will subtly influence how it works, and this will be specific to your age, sex and where in the world you find yourself. The microbiome plays a major part in explaining this phenomenon; your life is a unique book in the universal microbe library. Because we're still mapping so much of this biological complexity, surgery and interventional therapy remain the mainstay of treatment for multiple chronic diseases such as cancer – and I continue to take out appendixes at St Mary's Hospital (except that you'll now find me in theatre number seven).

Tomorrow we won't have the luxury of antibiotics as we now know them. Our addiction to these transformational medicines has fundamentally reshaped our internal biodiversity and impaired our resilience to both chronic and infectious diseases. The human 'resistome' is a biomarker of failure, and a change in strategy is not optional. We must reject abiosis as a life goal – the time has come to move towards a microbiome theory that accounts for our co-dependence on microbes in our growth, development and wellness, and that prioritizes microbial diversity and its core symbiotic functions.

This must promote socialization and inclusion, not the isolation and loneliness that come from putting up walls. Everyone should be able to make the most of the benefits their symbionts have to offer, and enjoy access to the next generation of vaccines and biotechnology that will permit the targeted eradication of pathogens.

Like parents everywhere, the Girl and I do the best we can to protect and nurture our children's microbiomes, but it's hard in a world of ubiquitous antibiotics, plastic, advertising, conflict and climate change. Reversing almost 200 years of microbiome mismanagement, caused by industrialization and globalization, can't be done through dietary changes and mindfulness alone. I'm sure you recycle and think about your carbon footprint, and you may already make choices about where you buy your foods and goods. These strategies will protect our microbiomes for future generations. However, there are some additional things you can do: don't buy foods from companies that don't have an antibiotic policy, from drug companies that promote antibiotic resistance or from those that compound our children's exposure to poor gut health.

Future health policy must acknowledge that an optimal parental microbiome is a prerequisite for health – this should be enshrined as a basic human right. Second, meaningful disease prevention must embrace a culture of *internal* climate activism. The evolutionary microbiome connects us to the environment in which we live and to all the other holobionts we share this planet with.

As a society, we need to empower schools, care homes and hospitals that care for people at critical moments of microbiome development, or failure, to make positive choices that promote gut health and optimal microbiome assembly and function. Our offices and homes (which are increasingly the same spaces) should also be designed to promote microbial diversity. The microbiome could be just the beginning of the revolution in preventative health.

Fritz Zwicky and Carl Woese were able to provide new insights through meticulous, precise measurements of our physical universe and, as is so often the case, scientific reality turned out to be so much stranger than fiction. We are still unable to see dark matter – one of the most important forces that govern all the known universe – yet we know it's there. The human microbiome is now undergoing a similar reinterpretation, and its importance to our existence is equally profound. The dark matter of the gut will not stay dark for long, because it offers a limitless new source of therapies that can be engineered to maintain not only our health, but also that of the planet. There is no doubt that the functions of the microbiome will turn out to be far weirder than we can currently imagine.

Until then, please don't make your own faecal transplant.

Notes

Prologue: Scrubbing Up

1 Lee S, Jang E J, Jo J, Park S J, Ryu H G. Long-term impacts of appendectomy associated with increased incidence of inflammatory bowel disease, infection, and colorectal cancer. *Int J Colorectal Dis* 2021; **36**(8): 1643–52.
2 Costs of the twenty-year war on terror: US $8 trillion and 900,000 deaths. Watson Institute for International and Public Affairs: Brown University, 2021.
3 Asher M I, Rutter C E, Bissell K, et al. Worldwide trends in the burden of asthma symptoms in school-aged children: Global Asthma Network Phase I cross-sectional study. *Lancet* 2021; **398**(10311): 1569–80.
4 Majeed H, Majeed H. Acknowledging the environmental allergy epidemic in children in the UK. *Lancet Planet Health* 2017; **1**(9): e349–e50.
5 GBD 2017 Inflammatory Bowel Disease Collaborators. The global, regional, and national burden of inflammatory bowel disease in 195 countries and territories, 1990–2017: a systematic analysis for the Global Burden of Disease Study 2017. *Lancet Gastroenterol Hepatol* 2020; **5**(1): 17–30.
6 Maenner M J, Shaw K A, Baio J, et al. Prevalence of Autism Spectrum Disorder Among Children Aged 8 Years – Autism and Developmental Disabilities Monitoring Network, 11 Sites, United States, 2016. *MMWR Surveill Summ* 2020; **69**(4): 1–12.
7 WHO. Obesity and overweight, 2021. www.who.int/news-room/fact-sheets/detail/obesity-and-overweight
8 Siegel R L, Fedewa S A, Anderson W F, et al. Colorectal Cancer Incidence Patterns in the United States, 1974–2013. *J Natl Cancer Inst* 2017; **109**(8).
9 International Asd. World Alzheimer Report 2015. The Global Impact of Dementia: An analysis of prevalence, incidence, cost and trends, 2015.

Introduction: The Faecal Microbiota Transplant

1 Smits W K, Lyras D, Lacy D B, Wilcox M H, Kuijper E J. Clostridium difficile infection. *Nat Rev Dis Primers* 2016; **2**: 16020.
2 van Nood E, Vrieze A, Nieuwdorp M, et al. Duodenal infusion of donor feces for recurrent Clostridium difficile. *N Engl J Med* 2013; **368**(5): 407–15.

3 El-Salhy M, Winkel R, Casen C, Hausken T, Gilja O H, Hatlebakk J G. Efficacy of Fecal Microbiota Transplantation for Patients with Irritable Bowel Syndrome at 3 Years After Transplantation. *Gastroenterology* 2022; **163**(4): 982–94.

1. A Library of the Known Microbial Universe

1 Teasdale M D, Fiddyment S, Vnoucek J, et al. The York Gospels: a 1000-year biological palimpsest. *R Soc Open Sci* 2017; **4**(10): 170988.
2 Hirsch E B, Raux B R, Lancaster J W, Mann R L, Leonard S N. Surface microbiology of the iPad tablet computer and the potential to serve as a fomite in both inpatient practice settings as well as outside of the hospital environment. *PLoS One* 2014; **9**(10): e111250.
3 Morell V. Microbiology's scarred revolutionary. *Science* 1997; **276**(5313): 699–702.
4 Gutell R R. You tell Carl that some of my best friends are Eukaryotes: Carl R. Woese (1928–2012). *RNA* 2013; **19**(4): vii–xi.
5 Baker B J, De Anda V, Seitz K W, Dombrowski N, Santoro A E, Lloyd K G. Diversity, ecology and evolution of Archaea. *Nat Microbiol* 2020; **5**(7): 887–900.
6 Venter J C, Remington K, Heidelberg J F, et al. Environmental genome shotgun sequencing of the Sargasso Sea. *Science* 2004; **304**(5667): 66–74.
7 DuPont H L. Giardia: both a harmless commensal and a devastating pathogen. *J Clin Invest* 2013; **123**(6): 2352–4.
8 Hallen-Adams H E, Suhr M J. Fungi in the healthy human gastrointestinal tract. *Virulence* 2017; **8**(3): 352–8.
9 Doron I, Leonardi I, Li X V, et al. Human gut mycobiota tune immunity via CARD9-dependent induction of anti-fungal IgG antibodies. *Cell* 2021; **184**(4): 1017–31 e14.
10 Nilsson R H, Anslan S, Bahram M, Wurzbacher C, Baldrian P, Tedersoo L. Mycobiome diversity: high-throughput sequencing and identification of fungi. *Nat Rev Microbiol* 2019; **17**(2): 95–109.
11 van Tilburg Bernardes E, Pettersen V K, Gutierrez M W, et al. Intestinal fungi are causally implicated in microbiome assembly and immune development in mice. *Nat Commun* 2020; **11**(1): 2577.
12 Microbiology by numbers. *Nat Rev Microbiol* 2011; **9**(9): 628.
13 Mushegian A R. Are There 10(31) Virus Particles on Earth, or More, or Fewer? *J Bacteriol* 2020; **202**(9).
14 Moreno-Gallego J L, Chou S P, Di Rienzi S C, et al. Virome Diversity Correlates with Intestinal Microbiome Diversity in Adult Monozygotic Twins. *Cell Host Microbe* 2019; **25**(2): 261–72 e5.

15 Camarillo-Guerrero L F, Almeida A, Rangel-Pineros G, Finn R D, Lawley T D. Massive expansion of human gut bacteriophage diversity. *Cell* 2021; **184**(4): 1098–9 e9.
16 Rohwer F. Global phage diversity. *Cell* 2003; **113**(2): 141.
17 Chen Z, Sun L, Zhang Z, et al. Cryo-EM structure of the bacteriophage T4 isometric head at 3.3-A resolution and its relevance to the assembly of icosahedral viruses. *Proc Natl Acad Sci U S A* 2017; **114**(39): E8184–E93.
18 Rao V B, Black L W. Structure and assembly of bacteriophage T4 head. *Virol J* 2010; **7**: 356.
19 Clokie M R, Millard A D, Letarov A V, Heaphy S. Phages in nature. *Bacteriophage* 2011; **1**(1): 31–45.
20 Breitbart M, Bonnain C, Malki K, Sawaya N A. Phage puppet masters of the marine microbial realm. *Nat Microbiol* 2018; **3**(7): 754–66.

2. Midlife Crisis

1 Hoch J A, Losick R. Panspermia, spores and the Bacillus subtilis genome. *Nature* 1997; **390**(6657): 237–8.
2 Horneck G. Survival of microorganisms in space: a review. *Adv Space Res* 1981; **1**(14): 39–48.
3 Leander B S. Predatory protists. *Curr Biol* 2020; **30**(10): R510–R16.
4 Sagan L. On the origin of mitosing cells. *J Theor Biol* 1967; **14**(3): 255–74.
5 Adams E D, Goss G G, Leys S P. Freshwater sponges have functional, sealing epithelia with high transepithelial resistance and negative transepithelial potential. *PLoS One* 2010; **5**(11): e15040.
6 Schiffbauer J D, Selly T, Jacquet S M, et al. Discovery of bilaterian-type through-guts in cloudinomorphs from the terminal Ediacaran Period. *Nat Commun* 2020; **11**(1): 205.
7 Flajnik M F, Kasahara M. Origin and evolution of the adaptive immune system: genetic events and selective pressures. *Nat Rev Genet* 2010; **11**(1): 47–59.
8 Nino Barreat J G, Katzourakis A. Evolutionary Analysis of Placental Orthologues Reveals Two Ancient DNA Virus Integrations. *J Virol* 2022: e0093322.
9 Long J A, Mark-Kurik E, Johanson Z, et al. Copulation in antiarch placoderms and the origin of gnathostome internal fertilization. *Nature* 2015; **517**(7533): 196–9.
10 Lively C M. A review of Red Queen models for the persistence of obligate sexual reproduction. *J Hered* 2010; **101** (Suppl 1): S13–20.
11 Poinar G, Zavortink T J, Brown A. Priscoculex burmanicus n. gen. et sp. (Diptera: Culicidae: Anophelinae) from mid-Cretaceous Myanmar amber. *Hist Biol* 2020; **32**(9): 1157–62.

12 Whitfield J. Portrait of a serial killer. *Nature* 2002.
13 Oftedal O T. The mammary gland and its origin during synapsid evolution. *J Mammary Gland Biol Neoplasia* 2002; 7(3): 225–52.
14 Oftedal O T. The origin of lactation as a water source for parchment-shelled eggs. *J Mammary Gland Biol Neoplasia* 2002; 7(3): 253–66.
15 Morono Y, Ito M, Hoshino T, et al. Aerobic microbial life persists in oxic marine sediment as old as 101.5 million years. *Nat Commun* 2020; 11(1): 3626.
16 Lugli G A, Milani C, Mancabelli L, et al. Ancient bacteria of the Otzi's microbiome: a genomic tale from the Copper Age. *Microbiome* 2017; 5(1): 5.
17 Moeller A H, Caro-Quintero A, Mjungu D, et al. Cospeciation of gut microbiota with hominids. *Science* 2016; 353(6297): 380–82.
18 Reese A T, Pereira F C, Schintlmeister A, et al. Microbial nitrogen limitation in the mammalian large intestine. *Nature Microbiology* 2018; 3(12): 1441–50.
19 Groussin M, Mazel F, Sanders J G, et al. Unraveling the processes shaping mammalian gut microbiomes over evolutionary time. *Nat Commun* 2017; 8: 14319.
20 Smith H F, Parker W, Kotzé S H, Laurin M. Multiple independent appearances of the cecal appendix in mammalian evolution and an investigation of related ecological and anatomical factors. *Comptes Rendus Palevol* 2013; 12(6): 339–54.
21 Weyrich L S, Duchene S, Soubrier J, et al. Neanderthal behaviour, diet, and disease inferred from ancient DNA in dental calculus. *Nature* 2017; 544(7650): 357–61.
22 Komiya Y, Shimomura Y, Higurashi T, et al. Patients with colorectal cancer have identical strains of Fusobacterium nucleatum in their colorectal cancer and oral cavity. *Gut* 2019; 68(7): 1335–7.
23 Wibowo M C, Yang Z, Borry M, et al. Reconstruction of ancient microbial genomes from the human gut. *Nature* 2021; 594(7862): 234–9.
24 Lloyd-Price J, Arze C, Ananthakrishnan A N, et al. Multi-omics of the gut microbial ecosystem in inflammatory bowel diseases. *Nature* 2019; 569(7758): 655–62.
25 Gomez A, Petrzelkova K J, Burns M B, et al. Gut Microbiome of Coexisting BaAka Pygmies and Bantu Reflects Gradients of Traditional Subsistence Patterns. *Cell Rep* 2016; 14(9): 2142–53.
26 Hansen M E B, Rubel M A, Bailey A G, et al. Population structure of human gut bacteria in a diverse cohort from rural Tanzania and Botswana. *Gen Biology* 2019; 20(1): 16.
27 Granato E T, Meiller-Legrand T A, Foster K R. The Evolution and Ecology of Bacterial Warfare. *Curr Biology* 2019; 29(11): R521–R37.
28 Czaran T L, Hoekstra R F, Pagie L. Chemical warfare between microbes promotes biodiversity. *Proc Natl Acad Sci U S A* 2002; 99(2): 786–90.

29. Niehus R, Oliveira N M, Li A, Fletcher A G, Foster K R. The evolution of strategy in bacterial warfare via the regulation of bacteriocins and antibiotics. *Elife* 2021; **10**.
30. Dunbar R I, Shultz S. Understanding primate brain evolution. *Philos Trans R Soc Lond B Biol Sci* 2007; **362**(1480): 649–58.
31. Raichle M E, Gusnard D A. Appraising the brain's energy budget. *Proc Natl Acad Sci U S A* 2002; **99**(16): 10237–9.
32. Obregon-Tito A J, Tito R Y, Metcalf J, et al. Subsistence strategies in traditional societies distinguish gut microbiomes. *Nat Commun* 2015; **6**: 6505.

3. How Microbes Kill Us

1. Aurna Li, et al. The excavation of the Neolithic site at Hamin Mangha in Horqin Left Middle Banner, Inner Mongolia in 2011. *Chinese Archaeology* 2014; **14**(1): 10–17.
2. Maritz J M, Sullivan S A, Prill R J, Aksoy E, Scheid P, Carlton J M. Filthy lucre: A metagenomic pilot study of microbes found on circulating currency in New York City. *PLoS One* 2017; **12**(4): e0175527.
3. Meyer C, Lohr C, Gronenborn D, Alt K W. The massacre mass grave of Schoneck-Kilianstadten reveals new insights into collective violence in Early Neolithic Central Europe. *Proc Natl Acad Sci U S A* 2015; **112**(36): 11217–22.
4. Stone A C, Wilbur A K, Buikstra J E, Roberts C A. Tuberculosis and leprosy in perspective. *Am J Phys Anthropol* 2009; **140** (Suppl 49): 66–94.
5. Shriner D, Rotimi C N. Whole-Genome-Sequence-Based Haplotypes Reveal Single Origin of the Sickle Allele during the Holocene Wet Phase. *Am J Hum Genet* 2018; **102**(4): 547–56.
6. van Doorn H R. Emerging infectious diseases. *Medicine* 2014; **42**(1): 60–63.
7. Frith J. History of Tuberculosis. Part 1 – Phthisis, consumption and the White Plague. *JMVH* 2014; **22**.
8. WHO. The top 10 causes of death, 2020. www.who.int/news-room/factsheets/detail/the-top-10-causes-of-death
9. Jones D S, Podolsky S H, Greene J A. The burden of disease and the changing task of medicine. *N Engl J Med* 2012; **366**(25): 2333–8.
10. WHO. The top 10 causes of death, 2020. www.who.int/news-room/factsheets/detail/the-top-10-causes-of-death
11. Warren J R, Marshall B. Unidentified curved bacilli on gastric epithelium in active chronic gastritis. *Lancet* 1983; **1**(8336): 1273–5.
12. Blaser M. Antibiotic overuse: Stop the killing of beneficial bacteria. *Nature* 2011; **476**(7361): 393–4.

13. Lebwohl B, Blaser M J, Ludvigsson J F, et al. Decreased risk of celiac disease in patients with Helicobacter pylori colonization. *Am J Epidemiol* 2013; **178**(12): 1721–30.
14. Barberis I, Bragazzi N L, Galluzzo L, Martini M. The history of tuberculosis: from the first historical records to the isolation of Koch's bacillus. *J Prev Med Hyg* 2017; **58**(1): E9–E12.
15. Scott A J, Alexander J L, Merrifield C A, et al. International Cancer Microbiome Consortium consensus statement on the role of the human microbiome in carcinogenesis. *Gut* 2019; **68**(9): 1624–32.
16. Chuang H L, Chiu C C, Lo C, et al. Circulating gut microbiota-related metabolites influence endothelium plaque lesion formation in ApoE knockout rats. *PLoS One* 2022; **17**(5): e0264934.
17. Libby P. The changing landscape of atherosclerosis. *Nature* 2021; **592**(7855): 524–33.
18. Luo T, Guo Z, Liu D, et al. Deficiency of PSRC1 accelerates atherosclerosis by increasing TMAO production via manipulating gut microbiota and flavin monooxygenase 3. *Gut Microbes* 2022; **14**(1): 2077602.
19. Visconti A, Le Roy C I, Rosa F, et al. Interplay between the human gut microbiome and host metabolism. *Nat Commun* 2019; **10**(1): 4505.

4. *Inflammation*

1. Vogt T C. How Many Years of Life Did the Fall of the Berlin Wall Add? A Projection of East German Life Expectancy. *Gerontology* 2013; **59**(3): 276–82.
2. Kramer U, Schmitz R, Ring J, Behrendt H. What can reunification of East and West Germany tell us about the cause of the allergy epidemic? *Clin Exp Allergy* 2015; **45**(1): 94–107.
3. Allergy UK. Allergy Prevalence: Useful facts and figures. 2022. www.allergyuk.org/about-allergy/statistics-and-figures/
4. Muraro A, Worm M, Alviani C, et al. EAACI guidelines: Anaphylaxis (2021 update). *Allergy* 2022; **77**(2): 357–77.
5. Correale J, Equiza T R. Regulatory B cells, helminths, and multiple sclerosis. *Methods Mol Biol* 2014; **1190**: 257–69.
6. Tanasescu R, Tench C R, Constantinescu C S, et al. Hookworm Treatment for Relapsing Multiple Sclerosis: A Randomized Double-Blinded Placebo-Controlled Trial. *JAMA Neurol* 2020; **77**(9): 1089–98.
7. Wolday D, Gebrecherkos T, Arefaine Z G, et al. Effect of co-infection with intestinal parasites on COVID-19 severity: A prospective observational cohort study. *EClinicalMedicine* 2021; **39**: 101054.

8 Elgazzar A, Eltaweel A, Youssef S A, Hany B, Hafez M, Moussa H. Efficacy and Safety of Ivermectin for Treatment and prophylaxis of COVID-19 Pandemic. 2020. www.researchsquare.com/article/rs-100956/v3
9 Strachan D P. Hay fever, hygiene, and household size. *BMJ* 1989; **299**(6710): 1259–60.
10 Rook G A, Adams V, Hunt J, Palmer R, Martinelli R, Brunet L R. Mycobacteria and other environmental organisms as immunomodulators for immunoregulatory disorders. *Springer Semin Immunopathol* 2004; **25**(3–4): 237–55.
11 Haahtela T, Holgate S, Pawankar R, et al. The biodiversity hypothesis and allergic disease: world allergy organization position statement. *World Allergy Organ J* 2013; **6**(1): 3.
12 Blaser M J, Falkow S. What are the consequences of the disappearing human microbiota? *Nat Rev Microbiol* 2009; **7**(12): 887–94.
13 Bao R, Hesser L A, He Z, Zhou X, Nadeau K C, Nagler C R. Fecal microbiome and metabolome differ in healthy and food-allergic twins. *J Clin Invest* 2021; **131**(2).
14 Stein M M, Hrusch C L, Gozdz J, et al. Innate Immunity and Asthma Risk in Amish and Hutterite Farm Children. *N Engl J Med* 2016; **375**(5): 411–21.
15 Kim J H, Kim K, Kim W. Gut microbiota restoration through fecal microbiota transplantation: a new atopic dermatitis therapy. *Exp Mol Med* 2021; **53**(5): 907–16.
16 Rachid R. Evaluating the Safety and Efficacy of Oral Encapsulated Fecal Microbiota Transplant in Peanut Allergic Patients, 2021. clinicaltrials.gov/ct2/show/NCT02960074

5. Sex and Bugs and Oestradiol

1 Johnson K V. Gut microbiome composition and diversity are related to human personality traits. *Hum Microb J* 2020; **15**: np.
2 Sharon G, Segal D, Ringo J M, Hefetz A, Zilber-Rosenberg I, Rosenberg E. Commensal bacteria play a role in mating preference of Drosophila melanogaster. *Proc Natl Acad Sci U S A* 2010; **107**(46): 20051–6.
3 Ehman K D, Scott M E. Female mice mate preferentially with non-parasitized males. *Parasitology* 2002; **125**(Pt 5): 461–6.
4 Boulet M, Charpentier M J, Drea C M. Decoding an olfactory mechanism of kin recognition and inbreeding avoidance in a primate. *BMC Evol Biol* 2009; **9**: 281.
5 Kavaliers M, Ossenkopp K P, Choleris E. Social neuroscience of disgust. *Genes Brain Behav* 2019; **18**(1): e12508.

6 Zhou W, Qi D, Swaisgood R R, et al. Symbiotic bacteria mediate volatile chemical signal synthesis in a large solitary mammal species. *ISME J* 2021; **15**(7): 2070–80.

7 Lam T H, Verzotto D, Brahma P, et al. Understanding the microbial basis of body odor in pre-pubescent children and teenagers. *Microbiome* 2018; **6**(1): 213.

8 McLaughlin J, Watterson S, Layton A M, Bjourson A J, Barnard E, McDowell A. Propionibacterium acnes and Acne Vulgaris: New Insights from the Integration of Population Genetic, Multi-Omic, Biochemical and Host-Microbe Studies. *Microorganisms* 2019; **7**(5).

9 Dreno B, Dagnelie M A, Khammari A, Corvec S. The Skin Microbiome: A New Actor in Inflammatory Acne. *Am J Clin Dermatol* 2020; **21** (Suppl 1): 18–24.

10 Gurtler A, Laurenz S. The impact of clinical nutrition on inflammatory skin diseases. *J Dtsch Dermatol Ges* 2022; **20**(2): 185–202.

11 Rainer B M, Thompson K G, Antonescu C, et al. Impact of lifestyle and demographics on the gut microbiota of acne patients and the response to minocycline. *J Dermatolog Treat* 2021; **32**(8): 934–5.

12 Huang Y, Liu L, Chen L, Zhou L, Xiong X, Deng Y. Gender-Specific Differences in Gut Microbiota Composition Associated with Microbial Metabolites for Patients with Acne Vulgaris. *Ann Dermatol* 2021; **33**(6): 531–40.

13 Farahani L, Tharakan T, Yap T, Ramsay J W, Jayasena C N, Minhas S. The semen microbiome and its impact on sperm function and male fertility: A systematic review and meta-analysis. *Andrology* 2021; **9**(1): 115–44.

14 Nelson D E, Dong Q, Van der Pol B, et al. Bacterial communities of the coronal sulcus and distal urethra of adolescent males. *PLoS One* 2012; **7**(5): e36298.

15 Ratten L K, Plummer E L, Murray G L, et al. Sex is associated with the persistence of non-optimal vaginal microbiota following treatment for bacterial vaginosis: a prospective cohort study. *BJOG* 2021; **128**(4): 756–67.

16 Malki K, Shapiro J W, Price T K, et al. Genomes of Gardnerella Strains Reveal an Abundance of Prophages within the Bladder Microbiome. *PLoS One* 2016; **11**(11): e0166757.

17 Grainha T, Jorge P, Alves D, Lopes S P, Pereira M O. Unraveling Pseudomonas aeruginosa and Candida albicans Communication in Coinfection Scenarios: Insights Through Network Analysis. *Front Cell Infect Microbiol* 2020; **10**: 550505.

18 Kort R, Caspers M, van de Graaf A, van Egmond W, Keijser B, Roeselers G. Shaping the oral microbiota through intimate kissing. *Microbiome* 2014; **2**: 41.

19 CDC. 1 in 5 people in the U.S. have a sexually transmitted infection, 2021. www.cdc.gov/nchhstp/newsroom/2021/2018-STI-incidence-prevalence-estimates.html

20 KFF. The Global HIV/AIDS Epidemic, 2021. www.kff.org/global-health-policy/fact-sheet/the-global-hivaids-epidemic/

21 Xie Y, Sun J, Wei L, et al. Altered gut microbiota correlate with different immune responses to HAART in HIV-infected individuals. *BMC Microbiol* 2021; **21**(1): 11.
22 Foessleitner P, Petricevic L, Boerger I, et al. HIV infection as a risk factor for vaginal dysbiosis, bacterial vaginosis, and candidosis in pregnancy: A matched case-control study. *Birth* 2021; **48**(1): 139–46.
23 Warrier V, Zhang X, Reed P, et al. Genetic correlates of phenotypic heterogeneity in autism. *Nat Genet* 2022; **54**: 1293–1304.
24 Willsey H R, Willsey A J, Wang B, State M W. Genomics, convergent neuroscience and progress in understanding autism spectrum disorder. *Nat Rev Neurosci* 2022; **23**(6): 323–41.
25 WHO. Adolescent and young adult health, 2021. www.who.int/news-room/fact-sheets/detail/adolescents-health-risks-and-solutions
26 Kageyama D, Narita S, Watanabe M. Insect Sex Determination Manipulated by Their Endosymbionts: Incidences, Mechanisms and Implications. *Insects* 2012; **3**(1): 161–99.
27 Weger B D, Gachon F. Microbiota and the clock: sexual dimorphism matters! *Aging (Albany NY)* 2019; **11**(12): 3893–4.
28 Thion M S, Low D, Silvin A, et al. Microbiome Influences Prenatal and Adult Microglia in a Sex-Specific Manner. *Cell* 2018; **172**(3): 500–16 e16.
29 Uzan-Yulzari A, Turta O, Belogolovski A, et al. Neonatal antibiotic exposure impairs child growth during the first six years of life by perturbing intestinal microbial colonization. *Nat Commun* 2021; **12**(1): 443.
30 Huttenhower C, Gevers D, Knight R, et al. Structure, function and diversity of the healthy human microbiome. *Nature* 2012; **486**(7402): 207–14.
31 Zhang X, Zhong H, Li Y, et al. Sex- and age-related trajectories of the adult human gut microbiota shared across populations of different ethnicities. *Nature Aging* 2021; **1**(1): 87–100.
32 Coombes Z, Yadav V, McCoubrey L E, et al. Progestogens Are Metabolized by the Gut Microbiota: Implications for Colonic Drug Delivery. *Pharmaceutics* 2020; **12**(8).
33 Hua X, Cao Y, Morgan D M, et al. Longitudinal analysis of the impact of oral contraceptive use on the gut microbiome. *J Med Microbiol* 2022; **71**(4).
34 Tayachew B, Vanden Brink H, Garcia-Reyes Y, et al. Combined Oral Contraceptive Treatment Does Not Alter the Gut Microbiome but Affects Amino Acid Metabolism in Sera of Obese Girls With Polycystic Ovary Syndrome. *Front Physiol* 2022; **13**: 887077.
35 Balle C, Konstantinus I N, Jaumdally S Z, et al. Hormonal contraception alters vaginal microbiota and cytokines in South African adolescents in a randomized trial. *Nat Commun* 2020; **11**(1): 5578.

36 Chen K L A, Liu X, Zhao Y C, et al. Long-Term Administration of Conjugated Estrogen and Bazedoxifene Decreased Murine Fecal β-Glucuronidase Activity Without Impacting Overall Microbiome Community. *Scientific Reports* 2018; **8**(1): 8166.

37 Li D, Liu R, Wang M, et al. 3beta-Hydroxysteroid dehydrogenase expressed by gut microbes degrades testosterone and is linked to depression in males. *Cell Host Microbe* 2022; **30**(3): 329–39 e5.

38 Liu S, Cao R, Liu L, et al. Correlation Between Gut Microbiota and Testosterone in Male Patients With Type 2 Diabetes Mellitus. *Front Endocrinol (Lausanne)* 2022; **13**: 836485.

39 United Nations. World Fertility and Family Planning 2020. www.un.org/development/desa/pd/sites/www.un.org.development.desa.pd/files/files/documents/2020/Aug/un_2020_worldfertilityfamilyplanning_highlights.pdf

40 Levine H, Jørgensen N, Martino-Andrade A, et al. Temporal trends in sperm count: a systematic review and meta-regression analysis. *Human Reproduction Update* 2017; **23**(6): 646–59.

41 Zhang T, Sun P, Geng Q, et al. Disrupted spermatogenesis in a metabolic syndrome model: the role of vitamin A metabolism in the gut–testis axis. *Gut* 2021; **71**(1): 78–87.

42 Yassour M, Jason E, Hogstrom L J, et al. Strain-Level Analysis of Mother-to-Child Bacterial Transmission during the First Few Months of Life. *Cell Host Microbe* 2018; **24**(1): 146–54 e4.

43 Koedooder R, Singer M, Schoenmakers S, et al. The vaginal microbiome as a predictor for outcome of in vitro fertilization with or without intracytoplasmic sperm injection: a prospective study. *Hum Reprod* 2019; **34**(6): 1042–54.

44 Campisciano G, Iebba V, Zito G, et al. Lactobacillus iners and gasseri, Prevotella bivia and HPV Belong to the Microbiological Signature Negatively Affecting Human Reproduction. *Microorganisms* 2020; **9**(1).

45 Hao Y, Feng Y, Yan X, et al. Gut microbiota-testis axis: FMT improves systemic and testicular micro-environment to increase semen quality in type 1 diabetes. *Molecular Medicine* 2022; **28**(1): 45.

6. The Big Bang

1 de Goffau M C, Lager S, Sovio U, et al. Human placenta has no microbiome but can contain potential pathogens. *Nature* 2019; **572**(7769): 329–34.

2 Yang H, Guo R, Li S, et al. Systematic analysis of gut microbiota in pregnant women and its correlations with individual heterogeneity. *NPJ Biofilms Microbiomes* 2020; **6**(1): 32.

3 Kok D E, Steegenga W T, McKay J A. Folate and epigenetics: why we should not forget bacterial biosynthesis. *Epigenomics* 2018; **10**(9): 1147–50.
4 Koren O, Goodrich J K, Cullender T C, et al. Host remodeling of the gut microbiome and metabolic changes during pregnancy. *Cell* 2012; **150**(3): 470–80.
5 Crusell M K W, Hansen T H, Nielsen T, et al. Gestational diabetes is associated with change in the gut microbiota composition in third trimester of pregnancy and postpartum. *Microbiome* 2018; **6**(1): 89.
6 Lv L J, Li S H, Li S C, et al. Early-Onset Preeclampsia Is Associated With Gut Microbial Alterations in Antepartum and Postpartum Women. *Front Cell Infect Microbiol* 2019; **9**: 224.
7 Chu D M, Ma J, Prince A L, Antony K M, Seferovic M D, Aagaard K M. Maturation of the infant microbiome community structure and function across multiple body sites and in relation to mode of delivery. *Nat Med* 2017; **23**(3): 314–26.
8 Morais J, Marques C, Teixeira D, et al. Extremely preterm neonates have more Lactobacillus in meconium than very preterm neonates – the in utero microbial colonization hypothesis. *Gut Microbes* 2020; **12**(1): 1785804.
9 Rackaityte E, Halkias J, Fukui E M, et al. Viable bacterial colonization is highly limited in the human intestine in utero. *Nat Med* 2020; **26**(4): 599–607.
10 Bi Y, Tu Y, Zhang N, et al. Multiomics analysis reveals the presence of a microbiome in the gut of fetal lambs. *Gut* 2021; **70**(5): 853–64.
11 Cobo T, Vergara A, Collado M C, et al. Characterization of vaginal microbiota in women with preterm labor with intra-amniotic inflammation. *Sci Rep* 2019; **9**(1): 18963.
12 Blostein F, Gelaye B, Sanchez S E, Williams M A, Foxman B. Vaginal microbiome diversity and preterm birth: results of a nested case-control study in Peru. *Ann Epidemiol* 2020; **41**: 28–34.
13 Dinsdale N K, Castano-Rodriguez N, Quinlivan J A, Mendz G L. Comparison of the Genital Microbiomes of Pregnant Aboriginal and Non-aboriginal Women. *Front Cell Infect Microbiol* 2020; **10**: 523764.
14 Stemming the global caesarean section epidemic. *Lancet* 2018; **392**(10155): 1279.
15 Sandall J, Tribe R M, Avery L, et al. Short-term and long-term effects of caesarean section on the health of women and children. *Lancet* 2018; **392**(10155): 1349–57.
16 Obermajer T, Grabnar I, Benedik E, et al. Microbes in Infant Gut Development: Placing Abundance Within Environmental, Clinical and Growth Parameters. *Sci Rep* 2017; **7**(1): 11230.
17 Diaz Heijtz R, Wang S, Anuar F, et al. Normal gut microbiota modulates brain development and behavior. *Proc Natl Acad Sci USA* 2011; **108**(7): 3047–52.

18　Chen Y. Global infant formula products market: estimations and forecasts for production and consumption China: Gira 2017.
19　Xu J, Lawley B, Wong G, et al. Ethnic diversity in infant gut microbiota is apparent before the introduction of complementary diets. *Gut Microbes* 2020; **11**(5): 1362–73.
20　Zheng W, Zhao W, Wu M, et al. Microbiota-targeted maternal antibodies protect neonates from enteric infection. *Nature* 2020; **577**(7791): 543–8.
21　Fehr K, Moossavi S, Sbihi H, et al. Breastmilk Feeding Practices Are Associated with the Co-Occurrence of Bacteria in Mothers' Milk and the Infant Gut: the CHILD Cohort Study. *Cell Host Microbe* 2020; **28**(2): 285–97 e4.

7. Symbiotic Sentience

1　Wang P, Su X, Mao Q, Liu Y. The surgical removal of a live tapeworm with an interesting pathologic finding most likely representing the migration path: a case report of cerebral sparganosis. *Clinics (Sao Paulo)* 2012; **67**(7): 849–51.
2　Tahoun A, Mahajan S, Paxton E, et al. Salmonella transforms follicle-associated epithelial cells into M cells to promote intestinal invasion. *Cell Host Microbe* 2012; **12**(5): 645–56.
3　World Stroke Association. Learn about stroke. 2022. www.world-stroke.org/world-stroke-day-campaign/why-stroke-matters/learn-about-stroke
4　Walker M, Uranga C, Levy S H, Kelly C, Edlund A. Thrombus-associated microbiota in acute ischemic stroke patients. *Surg Neurol Int* 2022; **13**: 247.
5　Schwedhelm E, von Lucadou M, Peine S, et al. Trimethyllysine, vascular risk factors and outcome in acute ischemic stroke (MARK-STROKE). *Amino Acids* 2021; **53**(4): 555–61.
6　Xu K, Gao X, Xia G, et al. Rapid gut dysbiosis induced by stroke exacerbates brain infarction in turn. *Gut* 2021; **70**(8).
7　Stanley D, Mason L J, Mackin K E, et al. Translocation and dissemination of commensal bacteria in post-stroke infection. *Nat Med* 2016; **22**(11): 1277–84.
8　Nelson J W, Phillips S C, Ganesh B P, Petrosino J F, Durgan D J, Bryan R M. The gut microbiome contributes to blood-brain barrier disruption in spontaneously hypertensive stroke prone rats. *FASEB J* 2021; **35**(2): e21201.
9　Benakis C, Brea D, Caballero S, et al. Commensal microbiota affects ischemic stroke outcome by regulating intestinal gammadelta T cells. *Nat Med* 2016; **22**(5): 516–23.
10　Fuertig R, Azzinnari D, Bergamini G, et al. Mouse chronic social stress increases blood and brain kynurenine pathway activity and fear behaviour: Both effects are reversed by inhibition of indoleamine 2,3-dioxygenase. *Brain Behav Immun* 2016; **54**: 59–72.

11 Kaelberer M M, Rupprecht L E, Liu W W, Weng P, Bohorquez D V. Neuropod Cells: The Emerging Biology of Gut-Brain Sensory Transduction. *Annu Rev Neurosci* 2020; **43**: 337–53.

12 Rao S, Schieber A M P, O'Connor C P, Leblanc M, Michel D, Ayres J S. Pathogen-Mediated Inhibition of Anorexia Promotes Host Survival and Transmission. *Cell* 2017; **168**(3): 503–16 e12.

13 Valles-Colomer M, Falony G, Darzi Y, et al. The neuroactive potential of the human gut microbiota in quality of life and depression. *Nat Microbiol* 2019; **4**(4): 623–32.

14 Kiraly D D, Walker D M, Calipari E S, et al. Alterations of the Host Microbiome Affect Behavioral Responses to Cocaine. *Sci Rep* 2016; **6**: 35455.

15 Needham B D, Funabashi M, Adame M D, et al. A gut-derived metabolite alters brain activity and anxiety behaviour in mice. *Nature* 2022; **602**(7898): 647–53.

16 Dunphy-Doherty F, O'Mahony S M, Peterson V L, et al. Post-weaning social isolation of rats leads to long-term disruption of the gut microbiota-immune-brain axis. *Brain Behav Immun* 2018; **68**: 261–73.

17 Roman-Urrestarazu A, van Kessel R, Allison C, Matthews F E, Brayne C, Baron-Cohen S. Association of Race/Ethnicity and Social Disadvantage with Autism Prevalence in 7 Million School Children in England. *JAMA Pediatrics* 2021; **175**(6): e210054-e.

18 van 't Hof M, Tisseur C, van Berckelear-Onnes I, et al. Age at autism spectrum disorder diagnosis: A systematic review and meta-analysis from 2012 to 2019. *Autism* 2021; **25**(4): 862–73.

19 Kanner L. Autistic disturbances of affective contact. *Nervous Child* 1943; **2**: 217–50.

20 Holingue C, Kalb L G, Musci R, et al. Characteristics of the autism spectrum disorder gastrointestinal and related behaviors inventory in children. *Autism Res* 2022; **15**(6): 1142–55.

21 Lou M, Cao A, Jin C, et al. Deviated and early unsustainable stunted development of gut microbiota in children with autism spectrum disorder. *Gut* 2022; **71**(8): 1588–99.

22 Bermudez-Martin P, Becker J A J, Caramello N, et al. The microbial metabolite p-Cresol induces autistic-like behaviors in mice by remodeling the gut microbiota. *Microbiome* 2021; **9**(1): 157.

23 Xiao L, Yan J, Yang T, et al. Fecal Microbiome Transplantation from Children with Autism Spectrum Disorder Modulates Tryptophan and Serotonergic Synapse Metabolism and Induces Altered Behaviors in Germ-Free Mice. *mSystems* 2021; **6**(2).

24 al-Haddad B J S, Jacobsson B, Chabra S, et al. Long-term Risk of Neuropsychiatric Disease After Exposure to Infection In Utero. *JAMA Psychiatry* 2019; **76**(6): 594–602.

25 Kim S, Kim H, Yim Y S, et al. Maternal gut bacteria promote neurodevelopmental abnormalities in mouse offspring. *Nature* 2017; **549**(7673): 528–32.
26 Agusti A, Campillo I, Balzano T, et al. Bacteroides uniformis CECT 7771 Modulates the Brain Reward Response to Reduce Binge Eating and Anxiety-Like Behavior in Rat. *Mol Neurobiol* 2021; **58**(10): 4959–79.
27 Han N D, Cheng J, Delannoy-Bruno O, et al. Microbial liberation of N-methylserotonin from orange fiber in gnotobiotic mice and humans. *Cell* 2022; **185**(14): 2495–509 e11.
28 Mortas H, Bilici S, Karakan T. The circadian disruption of night work alters gut microbiota consistent with elevated risk for future metabolic and gastrointestinal pathology. *Chronobiol Int* 2020; **37**(7): 1067–81.
29 Serger E, Luengo-Gutierrez L, Chadwick J S, et al. The gut metabolite indole-3 propionate promotes nerve regeneration and repair. *Nature* 2022; **607**(7919): 585–92.
30 Tu K W, Chiu M C, Lin W J, Hsueh Y P, Lin C C, Chou J Y. Possible impacts of the predominant Bacillus bacteria on the Ophiocordyceps unilateralis s. l. in its infected ant cadavers. *Sci Rep* 2021; **11**(1): 22695.
31 Public Health England. Prescribed medicines review: summary. 2020. www.gov.uk/government/publications/prescribed-medicines-review-report/prescribed-medicines-review-summary
32 Office for Health Improvement and Disparities. COVID-19 mental health and wellbeing surveillance: report. 2022. www.gov.uk/government/publications/covid-19-mental-health-and-wellbeing-surveillance-report

8. How to Live to One Hundred

1 The Wilde Medal and Lecture of the Manchester Literary and Philosophical Society. *Br Med J* 1901; **1**(2104): 1027–8.
2 Bloom D E, Chatterji S, Kowal P, et al. Macroeconomic implications of population ageing and selected policy responses. *Lancet* 2015; **385**(9968): 649–57.
3 Fontana L, Partridge L, Longo V D. Extending healthy life span – from yeast to humans. *Science* 2010; **328**(5976): 321–6.
4 Lee P G, Cigolle C, Blaum C. The co-occurrence of chronic diseases and geriatric syndromes: the health and retirement study. *Journal of the American Geriatrics Society* 2009; **57**(3): 511–16.
5 Deelen J, Beekman M, Capri M, Franceschi C, Slagboom P E. Identifying the genomic determinants of aging and longevity in human population studies: progress and challenges. *BioEssays: news and reviews in molecular, cellular and developmental biology* 2013; **35**(4): 386–96.

6 Lopez-Otin C, Kroemer G. Hallmarks of health. *Cell* 2021; **184**(7): 1929–39.
7 Bikov A, Szabo H, Piroska M, et al. Gut Microbiome in Patients with Obstructive Sleep Apnoea. *Applied Sciences* 2022; **12**(4): 2007.
8 Huffnagle G B, Dickson R P, Lukacs N W. The respiratory tract microbiome and lung inflammation: a two-way street. *Mucosal Immunol* 2017; **10**(2): 299–306.
9 Tanne J H. Don't vape, CDC says, as US lung disease epidemic grows. *BMJ* 2019; **366**: l5479.
10 WHO. Tobacco, 2022. www.who.int/news-room/fact-sheets/detail/tobacco
11 Sapkota A R, Berger S, Vogel T M. Human pathogens abundant in the bacterial metagenome of cigarettes. *Environ Health Perspect* 2010; **118**(3): 351–6.
12 Turek E M, Cox M J, Hunter M, et al. Airway microbial communities, smoking and asthma in a general population sample. *EBioMedicine* 2021; **71**: 103538.
13 Huang C, Shi G. Smoking and microbiome in oral, airway, gut and some systemic diseases. *J Transl Med* 2019; **17**(1): 225.
14 Collins L. Was Jeanne Calment the Oldest Person Who Ever Lived – or a Fraud? *New York Times*, 10 Feb 2020.
15 Biagi E, Candela M, Fairweather-Tait S, Franceschi C, Brigidi P. Aging of the human metaorganism: the microbial counterpart. *Age* 2012; **34**(1): 247–67.
16 Zwielehner J, Liszt K, Handschur M, Lassl C, Lapin A, Haslberger A G. Combined PCR-DGGE fingerprinting and quantitative-PCR indicates shifts in fecal population sizes and diversity of Bacteroides, bifidobacteria and Clostridium cluster IV in institutionalized elderly. *Exp Gerontol* 2009; **44**(6–7): 440–46.
17 Jeffery I B, Lynch D B, O'Toole P W. Composition and temporal stability of the gut microbiota in older persons. *The ISME Journal* 2016; **10**(1): 170–82.
18 Sato Y, Atarashi K, Plichta D R, et al. Novel bile acid biosynthetic pathways are enriched in the microbiome of centenarians. *Nature* 2021; **599**(7885): 456–64.
19 Zhang R, Hou A. Host-Microbe Interactions in Caenorhabditis elegans. *ISRN microbiology* 2013; **2013**: 356451.
20 Man A L, Gicheva N, Nicoletti C. The impact of ageing on the intestinal epithelial barrier and immune system. *Cellular immunology* 2014; **289**(1–2): 112–18.
21 Cevenini E, Monti D, Franceschi C. Inflamm-ageing. *Curr Opin Clin Nutr Metab Care* 2013; **16**(1): 14–20.
22 Kibe R, Kurihara S, Sakai Y, et al. Upregulation of colonic luminal polyamines produced by intestinal microbiota delays senescence in mice. *Sci Rep* 2014; **4**: 4548.
23 Gremer L, Scholzel D, Schenk C, et al. Fibril structure of amyloid-beta(1-42) by cryo-electron microscopy. *Science* 2017; **358**(6359): 116–19.
24 Liu P, Wu L, Peng G, et al. Altered microbiomes distinguish Alzheimer's disease from amnestic mild cognitive impairment and health in a Chinese cohort. *Brain Behav Immun* 2019; **80**: 633–43.

25 Harach T, Marungruang N, Duthilleul N, et al. Reduction of Abeta amyloid pathology in APPPS1 transgenic mice in the absence of gut microbiota. *Sci Rep* 2017; 7: 41802.
26 Chen C, Liao J, Xia Y, et al. Gut microbiota regulate Alzheimer's disease pathologies and cognitive disorders via PUFA-associated neuroinflammation. *Gut* 2022; 71(11): 2233–52.
27 Chorell E, Andersson E, Evans M L, et al. Bacterial Chaperones CsgE and CsgC Differentially Modulate Human alpha-Synuclein Amyloid Formation via Transient Contacts. *PLoS One* 2015; 10(10): e0140194.
28 Blacher E, Bashiardes S, Shapiro H, et al. Potential roles of gut microbiome and metabolites in modulating ALS in mice. *Nature* 2019; 572(7770): 474–80.

9. *The Global Microbiome*

1 Magnabosco C, Lin L H, Dong H, et al. The biomass and biodiversity of the continental subsurface. *Nat Geosci* 2018; 11(10): 707–17.
2 Cavicchioli R, Ripple W J, Timmis K N, et al. Scientists' warning to humanity: microorganisms and climate change. *Nat Rev Microbiol* 2019; 17(9): 569–86.
3 National Commission on the BP Deepwater Horizon Oil Spill and Offshore Drilling (US). Deep Water: The gulf oil disaster and the future of offshore drilling 2011. www.govinfo.gov/content/pkg/GPO-OILCOMMISSION/pdf/GPO-OILCOMMISSION.pdf
4 Hamdan L J, Salerno J L, Reed A, Joye S B, Damour M. The impact of the Deepwater Horizon blowout on historic shipwreck-associated sediment microbiomes in the northern Gulf of Mexico. *Sci Rep* 2018; 8(1): 9057.
5 Chernikova T N, Bargiela R, Toshchakov S V, et al. Hydrocarbon-Degrading Bacteria Alcanivorax and Marinobacter Associated With Microalgae Pavlova lutheri and Nannochloropsis oculata. *Front Microbiol* 2020; 11: 572931.
6 Campanale C, Massarelli C, Savino I, Locaputo V, Uricchio V F. A Detailed Review Study on Potential Effects of Microplastics and Additives of Concern on Human Health. *Int J Environ Res Public Health* 2020; 17(4).
7 Oliveri Conti G, Ferrante M, Banni M, et al. Micro- and nano-plastics in edible fruit and vegetables. The first diet risks assessment for the general population. *Environ Res* 2020; 187: 109677.
8 Qiao J, Chen R, Wang M, et al. Perturbation of gut microbiota plays an important role in micro/nanoplastics-induced gut barrier dysfunction. *Nanoscale* 2021; 13(19): 8806–16.

9 Tamargo A, Molinero N, Reinosa J J, et al. PET microplastics affect human gut microbiota communities during simulated gastrointestinal digestion, first evidence of plausible polymer biodegradation during human digestion. *Sci Rep* 2022; **12**(1): 528.
10 Wen S, Zhao Y, Liu S, Yuan H, You T, Xu H. Microplastics-perturbed gut microbiota triggered the testicular disorder in male mice: Via fecal microbiota transplantation. *Environ Pollut* 2022; **309**: 119789.
11 Yoshida S, Hiraga K, Takehana T, et al. A bacterium that degrades and assimilates poly(ethylene terephthalate). *Science* 2016; **351**(6278): 1196–9.
12 Elhacham E, Ben-Uri L, Grozovski J, Bar-On Y M, Milo R. Global human-made mass exceeds all living biomass. *Nature* 2020; **588**(7838): 442–4.
13 Liang Y, Zhan J, Liu D, et al. Organophosphorus pesticide chlorpyrifos intake promotes obesity and insulin resistance through impacting gut and gut microbiota. *Microbiome* 2019; **7**(1): 19.
14 Rothschild D, Weissbrod O, Barkan E, et al. Environment dominates over host genetics in shaping human gut microbiota. *Nature* 2018; **555**(7695): 210–15.
15 Goodrich J K, Davenport E R, Beaumont M, et al. Genetic Determinants of the Gut Microbiome in UK Twins. *Cell Host Microbe* 2016; **19**(5): 731–43.
16 Ou J, Carbonero F, Zoetendal E G, et al. Diet, microbiota, and microbial metabolites in colon cancer risk in rural Africans and African Americans. *Am J Clin Nutr* 2013; **98**(1): 111–20.
17 Truong D T, Tett A, Pasolli E, Huttenhower C, Segata N. Microbial strain-level population structure and genetic diversity from metagenomes. *Genome Res* 2017; **27**(4): 626–38.
18 Wibowo M C, Yang Z, Borry M, et al. Reconstruction of ancient microbial genomes from the human gut. *Nature* 2021; **594**(7862): 234–9.
19 Yatsunenko T, Rey F E, Manary M J, et al. Human gut microbiome viewed across age and geography. *Nature* 2012; **486**(7402): 222–7.
20 Smits S A, Leach J, Sonnenburg E D, et al. Seasonal cycling in the gut microbiome of the Hadza hunter-gatherers of Tanzania. *Science* 2017; **357**(6353): 802–6.
21 Hansen M E B, Rubel M A, Bailey A G, et al. Population structure of human gut bacteria in a diverse cohort from rural Tanzania and Botswana. *Genome Biol* 2019; **20**(1): 16.
22 Pasolli E, Asnicar F, Manara S, et al. Extensive Unexplored Human Microbiome Diversity Revealed by Over 150,000 Genomes from Metagenomes Spanning Age, Geography, and Lifestyle. *Cell* 2019; **176**(3): 649–62 e20.
23 Tamburini F B, Maghini D, Oduaran O H, et al. Short- and long-read metagenomics of urban and rural South African gut microbiomes reveal a transitional composition and undescribed taxa. *Nat Commun* 2022; **13**(1): 926.

24 United Nations News: www.un.org/development/desa/en/news/population/2018-revision-of-world-urbanization-prospects.html
25 Ruiz-Calderon J F, Cavallin H, Song S J, et al. Walls talk: Microbial biogeography of homes spanning urbanization. *Sci Adv* 2016; **2**(2): e1501061.
26 Danko D, Bezdan D, Afshin E E, et al. A global metagenomic map of urban microbiomes and antimicrobial resistance. *Cell* 2021; **184**(13): 3376–93 e17.
27 Lax S, Smith D P, Hampton-Marcell J, et al. Longitudinal analysis of microbial interaction between humans and the indoor environment. *Science* 2014; **345**(6200): 1048–52.
28 de Clercq N C, Frissen M N, Levin E, et al. The effect of having Christmas dinner with in-laws on gut microbiota composition. *Human Microbiome Journal* 2019; **13**: 100058.
29 Brito I L, Yilmaz S, Huang K, et al. Mobile genes in the human microbiome are structured from global to individual scales. *Nature* 2016; **535**(7612): 435–9.
30 Brand E C, Klaassen M A Y, Gacesa R, et al. Healthy Cotwins Share Gut Microbiome Signatures With Their Inflammatory Bowel Disease Twins and Unrelated Patients. *Gastroenterology* 2021; **160**(6): 1970–85.
31 International Organization for Migration. World Migration Report 2020. worldmigrationreport.iom.int/wmr-2020-interactive/
32 Maskarinec G, Noh J J. The effect of migration on cancer incidence among Japanese in Hawaii. *Ethn Dis* 2004; **14**(3): 431–9.

10. The War on Bugs

1 Murray C K, Hinkle M K, Yun H C. History of infections associated with combat-related injuries. *J Trauma* 2008; **64** (3 Suppl): S221–31.
2 Murray C K. Epidemiology of Infections Associated With Combat-Related Injuries in Iraq and Afghanistan. *J Trauma Acute Care Surg* 2008; **64**(3).
3 Bentley R. Different roads to discovery; Prontosil (hence sulfa drugs) and penicillin (hence β-lactams). *J Ind Microbiol Biotechnol* 2009; **36**(6): 775–86.
4 Discovery and Development of Penicillin. The National Historic Chemical Landmarks program of the American Chemical Society, 1999. www.acs.org/education/whatischemistry/landmarks/flemingpenicillin.html
5 Summers W C. Cholera and plague in India: the bacteriophage inquiry of 1927–1936. *J Hist Med Allied Sci* 1993; **48**(3): 275–301.
6 Newton P N, Timmermann B. Fake penicillin, *The Third Man*, and Operation Claptrap. *BMJ* 2016; **355**: i6494.
7 Acinetobacter baumannii infections among patients at military medical facilities treating injured U.S. service members, 2002–2004. *MMWR Morb Mortal Wkly Rep* 2004; **53**(45): 1063–6.

8 Johnstone P, Matheson A S. An invisible enemy: Panton-Valentine leukocidin Staphylococcus aureus on deployed troops. *J R Nav Med Serv* 2013; **99**(1): 9–12.

9 Alhasson F, Das S, Seth R, et al. Altered gut microbiome in a mouse model of Gulf War Illness causes neuroinflammation and intestinal injury via leaky gut and TLR4 activation. *PLoS One* 2017; **12**(3): e0172914.

10 Bose D, Saha P, Mondal A, et al. Obesity Worsens Gulf War Illness Symptom Persistence Pathology by Linking Altered Gut Microbiome Species to Long-Term Gastrointestinal, Hepatic, and Neuronal Inflammation in a Mouse Model. *Nutrients* 2020; **12**(9).

11 Koch A, Brierley C, Maslin M M, Lewis S L. Earth system impacts of the European arrival and Great Dying in the Americas after 1492. *Quat Sci Rev* 2019; **207**: 13–36.

12 Linden O, Jerneloev A, Egerup J. The Environmental Impacts of the Gulf War 1991: International Institute for Applied Systems Analysis, 2004.

13 UNHCR, Figures at a Glance, 2022. www.unhcr.org/uk/figures-at-a-glance.html

14 Häsler R, Kautz C, Rehman A, et al. The antibiotic resistome and microbiota landscape of refugees from Syria, Iraq and Afghanistan in Germany. *Microbiome* 2018; **6**(1): 37.

15 The National Archives. Conclusions of a Meeting of the War Cabinet held in the Cabinet War Room on Wednesday, 14th March, 1945, at 12 noon.

16 Landecker H. Antimicrobials before antibiotics: war, peace, and disinfectants. *Palgrave Commun* 2019; **5**(1): 45.

17 Moore P R, Evenson A, et al. Use of sulfasuxidine, streptothricin, and streptomycin in nutritional studies with the chick. *J Biol Chem* 1946; **165**(2): 437–41.

18 Siegel P B. Evolution of the modern broiler and feed efficiency. *Annu Rev Anim Biosci* 2014; **2**: 375–85.

19 Kirchhelle C. Toxic confusion: the dilemma of antibiotic regulation in West German food production (1951–1990). *Endeavour* 2016; **40**(2): 114–27.

20 Woods A. Science, disease and dairy production in Britain c.1927 to 1980. *AgHR* ; **62**(II): 294–314.

21 Kirchhelle C. Pharming animals: a global history of antibiotics in food production (1935–2017). *Palgrave Commun* 2018; **4**(1): 96.

22 Lulijwa R, Rupia E, Alfaro A. Antibiotic use in aquaculture, policies and regulation, health and environmental risks: a review of the top 15 major producers. *Rev Aquac* 2019; **12**.

23 Newey S. 'Unprecedented' bird flu outbreak exacerbating food insecurity. 2022. www.telegraph.co.uk/global-health/science-and-disease/unprecedented-bird-flu-outbreak-exacerbating-food-insecurity/

24 Malan-Muller S, Valles-Colomer M, Foxx C L, et al. Exploring the relationship between the gut microbiome and mental health outcomes in a posttraumatic stress disorder cohort relative to trauma-exposed controls. *Eur Neuropsychopharmacol* 2022; **56**: 24–38.

11. A Biotic Life

1. Essa A, Walfisch A, Sheiner E, Sergienko R, Wainstock T. Delivery mode and future infectious morbidity of the offspring: a sibling analysis. *Arch Gynecol Obstet* 2020; **302**(5): 1135–41.
2. Mandal M D, Mandal S. Honey: its medicinal property and antibacterial activity. *Asian Pac J Trop Biomed* 2011; **1**(2): 154–60.
3. Wang J, Xu C, Wong Y K, et al. Artemisinin, the Magic Drug Discovered from Traditional Chinese Medicine. *Engineering* 2019; **5**(1): 32–9.
4. Hall B G, Barlow M. Evolution of the serine beta-lactamases: past, present and future. *Drug Resist Updat* 2004; **7**(2): 111–23.
5. Santiago-Rodriguez T M, Fornaciari G, Luciani S, et al. Gut Microbiome of an 11th Century A.D. Pre-Columbian Andean Mummy. *PloS one* 2015; **10**(9): e0138135.
6. Starr M P, Reynolds D M. Streptomycin resistance of coliform bacteria from turkeys fed streptomycin. *Am J Public Health Nations Health* 1951; **41**(11 Pt 1): 1375–80.
7. Sales of veterinary antimicrobial agents in 31 European countries in 2019 and 2020: Trends from 2010 to 2020 Eleventh ESVAC report. European Medicines Agency, 2020.
8. Klein E Y, Van Boeckel T P, Martinez E M, et al. Global increase and geographic convergence in antibiotic consumption between 2000 and 2015. *Proc Natl Acad Sci U S A* 2018; **115**(15): E3463–E70.
9. Lubbert C, Baars C, Dayakar A, et al. Environmental pollution with antimicrobial agents from bulk drug manufacturing industries in Hyderabad, South India, is associated with dissemination of extended-spectrum beta-lactamase and carbapenemase-producing pathogens. *Infection* 2017; **45**(4): 479–91.
10. Kraupner N, Hutinel M, Schumacher K, et al. Evidence for selection of multi-resistant E. coli by hospital effluent. *Environ Int* 2021; **150**: 106436.
11. Public Health England. Health matters: antimicrobial resistance. 2015. www.gov.uk/government/publications/health-matters-antimicrobial-resistance
12. Clardy J, Fischbach M A, Currie C R. The natural history of antibiotics. *Cur Biol* 2009; **19**(11): R437–R41.

13 Nguyen L H, Ortqvist A K, Cao Y, et al. Antibiotic use and the development of inflammatory bowel disease: a national case-control study in Sweden. *Lancet Gastroenterol Hepatol* 2020; **5**(11): 986–95.

14 Wan Q Y, Zhao R, Wang Y, Wu Y, Wu X T. Antibiotic use and risk of colorectal cancer: a meta-analysis of 412 450 participants. *Gut* 2020; **69**(11): 2059–60.

15 Song M, Nguyen L H, Emilsson L, Chan A T, Ludvigsson J F. Antibiotic Use Associated with Risk of Colorectal Polyps in a Nationwide Study. *Clin Gastroenterol Hepatol* 2021; **19**(7): 1426-35 e6.

16 Yang B, Hagberg K W, Chen J, et al. Associations of antibiotic use with risk of primary liver cancer in the Clinical Practice Research Datalink. *Br J Cancer* 2016; **115**(1): 85–9.

17 Velicer C M, Heckbert S R, Lampe J W, Potter J D, Robertson C A, Taplin S H. Antibiotic use in relation to the risk of breast cancer. *JAMA* 2004; **291**(7): 827–35.

18 Hagan T, Cortese M, Rouphael N, et al. Antibiotics-Driven Gut Microbiome Perturbation Alters Immunity to Vaccines in Humans. *Cell* 2019; **178**(6): 1313–28.e13.

19 Zarrinpar A, Chaix A, Xu Z Z, et al. Antibiotic-induced microbiome depletion alters metabolic homeostasis by affecting gut signaling and colonic metabolism. *Nat Commun* 2018; **9**(1): 2872.

20 Anthony W E, Wang B, Sukhum K V, et al. Acute and persistent effects of commonly used antibiotics on the gut microbiome and resistome in healthy adults. *Cell Reports* 2022; **39**(2): 110649.

21 Torres-Barceló C, Gurney J, Gougat-Barberá C, Vasse M, Hochberg M E. Transient negative effects of antibiotics on phages do not jeopardise the advantages of combination therapies. *FEMS Microbiol Ecol* 2018; **94**(8).

22 Rogawski E T, Platts-Mills J A, Seidman J C, et al. Use of antibiotics in children younger than two years in eight countries: a prospective cohort study. *Bull World Health Organ* 2017; **95**(1): 49–61.

23 CDC. Antibiotic Use in the United States, 2020. Update: Progress and Opportunities. www.cdc.gov/antibiotic-use/stewardship-report/2020.html

24 Tzialla C, Borghesi A, Perotti G F, Garofoli F, Manzoni P, Stronati M. Use and misuse of antibiotics in the neonatal intensive care unit. *J Matern Fetal Neonatal Med* 2012; **25** (Suppl 4): 35–7.

25 Zhong Y, Zhang Y, Wang Y, Huang R. Maternal antibiotic exposure during pregnancy and the risk of allergic diseases in childhood: A meta-analysis. *Pediatr Allergy Immunol* 2021; **32**(3): 445–56.

26 Zhang Z, Wang J, Wang H, et al. Association of infant antibiotic exposure and risk of childhood asthma: A meta-analysis. *The World Allergy Organization Journal* 2021; **14**(11): 100607.

27 Sultan A A, Mallen C, Muller S, et al. Antibiotic use and the risk of rheumatoid arthritis: a population-based case-control study. *BMC Med* 2019; **17**(1): 154.
28 Fan H, Gilbert R, O'Callaghan F, Li L. Associations between macrolide antibiotics prescribing during pregnancy and adverse child outcomes in the UK: population based cohort study. *BMJ* 2020; **368**: m331.
29 Korpela K, Salonen A, Virta L J, et al. Intestinal microbiome is related to lifetime antibiotic use in Finnish pre-school children. *Nat Commun* 2016; **7**: 10410.
30 Oliveira R A, Cabral V, Xavier K B. Microbiome-diet interactions drive antibiotic efficacy. *Nat Microbiol* 2021; **6**(7): 824–5.
31 Korpela K, Zijlmans M A, Kuitunen M, et al. Childhood BMI in relation to microbiota in infancy and lifetime antibiotic use. *Microbiome* 2017; **5**(1): 26.
32 WHO. Immunization, 2019. www.who.int/news-room/facts-in-pictures/detail/immunization
33 Lewnard J A, Lo N C, Arinaminpathy N, Frost I, Laxminarayan R. Childhood vaccines and antibiotic use in low- and middle-income countries. *Nature* 2020; **581**(7806): 94–9.
34 Lynn M A, Tumes D J, Choo J M, et al. Early-Life Antibiotic-Driven Dysbiosis Leads to Dysregulated Vaccine Immune Responses in Mice. *Cell Host Microbe* 2018; **23**(5): 653–60.e5.
35 Roy A, Eisenhut M, Harris R J, et al. Effect of BCG vaccination against Mycobacterium tuberculosis infection in children: systematic review and meta-analysis. *BMJ* 2014; **349**: g4643.

12. The Drugs Don't Work

1 Global Medicines Use in 2020. IMS institute for healthcare informatics, 2021.
2 Global Medicine Spending and Usage Trends: outlook to 2025. IQVIA institute, 2021.
3 Wouters O J, McKee M, Luyten J. Estimated Research and Development Investment Needed to Bring a New Medicine to Market, 2009-2018. *Jama* 2020; **323**(9): 844–53.
4 Wide-ranging online data for epidemiologic research (WONDER). Atlanta, GA: CDC, National Center for Health Statistics. 2020. wonder.cdc.gov
5 United States Bankruptcy Court, Southern District of New York. Mediator's Fourth Interim Report, 2022.
6 Xu Y, Xie Z, Wang H, et al. Bacterial Diversity of Intestinal Microbiota in Patients with Substance Use Disorders Revealed by 16S rRNA Gene Deep Sequencing. *Scientific Reports* 2017; **7**(1): 3628.

7 Zhang J, Deji C, Fan J, et al. Differential alteration in gut microbiome profiles during acquisition, extinction and reinstatement of morphine-induced CPP. *Prog Neuropsychopharmacol Biol Psychiatry* 2021; **104**: 110058.
8 Zhang L, Meng J, Ban Y, et al. Morphine tolerance is attenuated in germfree mice and reversed by probiotics, implicating the role of gut microbiome. *Proc Natl Acad Sci U S A* 2019; **116**(27): 13523–32.
9 Simpson S, Kimbrough A, Boomhower B, et al. Depletion of the Microbiome Alters the Recruitment of Neuronal Ensembles of Oxycodone Intoxication and Withdrawal. *eneuro* 2020; **7**(3): ENEURO.0312–19.2020.
10 Lee K, Vuong H E, Nusbaum D J, Hsiao E Y, Evans C J, Taylor A M W. The gut microbiota mediates reward and sensory responses associated with regimen-selective morphine dependence. *Neuropsychopharmacology* 2018; **43**(13): 2606–14.
11 El-Salhy M, Winkel R, Casen C, Hausken T, Gilja O H, Hatlebakk J G. Efficacy of Fecal Microbiota Transplantation for Patients With Irritable Bowel Syndrome at 3 Years After Transplantation. *Gastroenterology* 2022; **163**(4): 982–94.e14.
12 Shen S, Lim G, You Z, et al. Gut microbiota is critical for the induction of chemotherapy-induced pain. *Nat Neurosci* 2017; **20**(9): 1213–16.
13 Rebello D, Wang E, Yen E, Lio P A, Kelly C R. Hair Growth in Two Alopecia Patients after Fecal Microbiota Transplant. *ACG Case Rep J* 2017; **4**: e107-e.
14 Yu T, Guo F, Yu Y, et al. Fusobacterium nucleatum Promotes Chemoresistance to Colorectal Cancer by Modulating Autophagy. *Cell* 2017; **170**(3): 548–63.e16.
15 Sivan A, Corrales L, Hubert N, et al. Commensal Bifidobacterium promotes antitumor immunity and facilitates anti-PD-L1 efficacy. *Science (New York, NY)* 2015; **350**(6264): 1084–9.
16 Abrahami D, McDonald E G, Schnitzer M E, Barkun A N, Suissa S, Azoulay L. Proton pump inhibitors and risk of gastric cancer: population-based cohort study. *Gut* 2022; **71**(1): 16–24.
17 Xia B, Yang M, Nguyen L H, et al. Regular Use of Proton Pump Inhibitor and the Risk of Inflammatory Bowel Disease: Pooled Analysis of 3 Prospective Cohorts. *Gastroenterology* 2021; **161**(6): 1842–52 e10.
18 Moledina D G, Perazella M A. Proton Pump Inhibitors and CKD. *J Am Soc Nephrol* 2016; **27**(10): 2926–8.
19 Yang M, He Q, Gao F, et al. Regular use of proton-pump inhibitors and risk of stroke: a population-based cohort study and meta-analysis of randomized-controlled trials. *BMC Med* 2021; **19**(1): 316.
20 Torres-Bondia F, Dakterzada F, Galván L, et al. Proton pump inhibitors and the risk of Alzheimer's disease and non-Alzheimer's dementias. *Scientific Reports* 2020; **10**(1): 21046.

13. The Microbiome Café

1. Hutchinson C. Concluding remarks, Coldspring Harbor Symposium. *Quant Biol* 1957; **22**: 415–27.
2. Zuo T, Zhang F, Lui G C Y, et al. Alterations in Gut Microbiota of Patients with COVID-19 During Time of Hospitalization. *Gastroenterology* 2020; **159**(3): 944–55.e8.
3. The 2021 Uber Eats Cravings Report. 2021. www.uber.com/newsroom/the-2021-uber-eats-cravings-report/
4. Zarrinpar A, Chaix A, Xu Z Z, et al. Antibiotic-induced microbiome depletion alters metabolic homeostasis by affecting gut signaling and colonic metabolism. *Nat Commun* 2018; **9**(1): 2872.
5. Moran-Ramos S, Macias-Kauffer L, López-Contreras B E, et al. A higher bacterial inward BCAA transport driven by Faecalibacterium prausnitzii is associated with lower serum levels of BCAA in early adolescents. *Mol Med* 2021; **27**(1): 108.
6. Jung M J, Lee J, Shin N R, et al. Chronic Repression of mTOR Complex 2 Induces Changes in the Gut Microbiota of Diet-induced Obese Mice. *Sci Rep* 2016; **6**: 30887.
7. Kenny D J, Plichta D R, Shungin D, et al. Cholesterol Metabolism by Uncultured Human Gut Bacteria Influences Host Cholesterol Level. *Cell Host Microbe* 2020; **28**(2): 245–57.e6.
8. Seidelmann S B, Claggett B, Cheng S, et al. Dietary carbohydrate intake and mortality: a prospective cohort study and meta-analysis. *The Lancet Public Health* 2018; **3**(9): e419–e28.
9. Miesen P, van Rij R P. Crossing the Mucosal Barrier: A Commensal Bacterium Gives Dengue Virus a Leg-Up in the Mosquito Midgut. *Cell Host Microbe* 2019; **25**(1): 1–2.
10. Liu R, Hong J, Xu X, et al. Gut microbiome and serum metabolome alterations in obesity and after weight-loss intervention. *Nat Med* 2017; **23**(7): 859–68.
11. Tan H-E, Sisti A C, Jin H, et al. The gut–brain axis mediates sugar preference. *Nature* 2020; **580**(7804): 511–16.
12. Ervin B R, Ogden C L. Consumption of Added Sugars Among U.S. Adults, 2005–2010, NCHS Data Brief, 2013.
13. Suez J, Korem T, Zeevi D, et al. Artificial sweeteners induce glucose intolerance by altering the gut microbiota. *Nature* 2014; **514**(7521): 181–6.
14. O'Keefe S J, Li J V, Lahti L, et al. Fat, fibre and cancer risk in African Americans and rural Africans. *Nat Commun* 2015; **6**: 6342.
15. Katsirma Z, Dimidi E, Rodriguez-Mateos A, Whelan K. Fruits and their impact on the gut microbiota, gut motility and constipation. *Food Funct* 2021; **12**(19): 8850–66.

16 Wastyk H C, Fragiadakis G K, Perelman D, et al. Gut-microbiota-targeted diets modulate human immune status. *Cell* 2021; **184**(16): 4137–53.e14.
17 Lin H C, Peng C H, Huang C N, Chiou J Y. Soy-Based Foods Are Negatively Associated with Cognitive Decline in Taiwan's Elderly. *J Nutr Sci Vitaminol (Tokyo)* 2018; **64**(5): 335–9.
18 Rowland I R, Wiseman H, Sanders T A, Adlercreutz H, Bowey E A. Interindividual variation in metabolism of soy isoflavones and lignans: influence of habitual diet on equol production by the gut microflora. *Nutr Cancer* 2000; **36**(1): 27–32.
19 Iino C, Shimoyama T, Iino K, et al. Daidzein Intake Is Associated with Equol Producing Status through an Increase in the Intestinal Bacteria Responsible for Equol Production. *Nutrients* 2019; **11**(2): 433.
20 Chassaing B, Koren O, Goodrich J K, et al. Dietary emulsifiers impact the mouse gut microbiota promoting colitis and metabolic syndrome. *Nature* 2015; **519**(7541): 92–6.
21 Naimi S, Viennois E, Gewirtz A T, Chassaing B. Direct impact of commonly used dietary emulsifiers on human gut microbiota. *Microbiome* 2021; **9**(1): 66.
22 Ostrowski M P, La Rosa S L, Kunath B J, et al. Mechanistic insights into consumption of the food additive xanthan gum by the human gut microbiota. *Nat Microbiol* 2022; **7**(4): 556–69.
23 Carmody R N, Bisanz J E, Bowen B P, et al. Cooking shapes the structure and function of the gut microbiome. *Nature Microbiology* 2019; **4**(12): 2052–63.
24 Diederen K, Li J V, Donachie G E, et al. Exclusive enteral nutrition mediates gut microbial and metabolic changes that are associated with remission in children with Crohn's disease. *Sci Rep* 2020; **10**(1): 18879.
25 Le Roy C I, Wells P M, Si J, Raes J, Bell J T, Spector T D. Red Wine Consumption Associated with Increased Gut Microbiota α-Diversity in 3 Independent Cohorts. *Gastroenterology* 2020; **158**(1): 270–72.e2.
26 Leclercq S, Matamoros S, Cani P D, et al. Intestinal permeability, gut-bacterial dysbiosis, and behavioral markers of alcohol-dependence severity. *Proc Natl Acad Sci U S A* 2014; **111**(42): E4485–93.
27 Piacentino D, Grant-Beurmann S, Vizioli C, et al. Gut microbiome and metabolome in a non-human primate model of chronic excessive alcohol drinking. *Transl Psychiatry* 2021; **11**(1): 609.
28 Duan Y, Llorente C, Lang S, et al. Bacteriophage targeting of gut bacterium attenuates alcoholic liver disease. *Nature* 2019; **575**(7783): 505–11.
29 LaMotte S. Woman claims her body brews alcohol, has DUI charge dismissed. 2016. edition.cnn.com/2015/12/31/health/auto-brewery-syndrome-dui-womans-body-brews-own-alcohol/index.html
30 Kaji H, Asanuma Y, Yahara O, et al. Intragastrointestinal alcohol fermentation syndrome: report of two cases and review of the literature. *J Forensic Sci Soc* 1984; **24**(5): 461–71.

31 Singh G M, Micha R, Khatibzadeh S, Lim S, Ezzati M, Mozaffarian D. Estimated Global, Regional, and National Disease Burdens Related to Sugar-Sweetened Beverage Consumption in 2010. *Circulation* 2015; **132**(8): 639–66.

14. Extreme Phenotypes

1 WHO. Obesity and overweight, 2021. www.who.int/news-room/fact-sheets/detail/obesity-and-overweight
2 McCarthy H, Potts H W W, Fisher A. Physical Activity Behavior Before, During, and After COVID-19 Restrictions: Longitudinal Smartphone-Tracking Study of Adults in the United Kingdom. *J Med Internet Res* 2021; **23**(2): e23701.
3 COVID-19 and Obesity: The 2021 Atlas: The cost of not addressing the global obesity crisis. worldobesity.org 2921.
4 Perino A, Velázquez-Villegas L A, Bresciani N, et al. Central anorexigenic actions of bile acids are mediated by TGR5. *Nat Metab* 2021; **3**(5): 595–603.
5 Psichas A, Sleeth M L, Murphy K G, et al. The short chain fatty acid propionate stimulates GLP-1 and PYY secretion via free fatty acid receptor 2 in rodents. *Int J Obes* 2015; **39**(3): 424–9.
6 Vatanen T, Kostic A D, d'Hennezel E, et al. Variation in Microbiome LPS Immunogenicity Contributes to Autoimmunity in Humans. *Cell* 2016; **165**(4): 842–53.
7 Rouland M, Beaudoin L, Rouxel O, et al. Gut mucosa alterations and loss of segmented filamentous bacteria in type 1 diabetes are associated with inflammation rather than hyperglycaemia. *Gut* 2022; **71**(2): 296–308.
8 Vatanen T, Franzosa E A, Schwager R, et al. The human gut microbiome in early-onset type 1 diabetes from the TEDDY study. *Nature* 2018; **562**(7728): 589–94.
9 Vehik K, Lynch K F, Wong M C, et al. Prospective virome analyses in young children at increased genetic risk for type 1 diabetes. *Nat Med* 2019; **25**(12): 1865–72.
10 International Diabetes Federation. About Diabetes, 2021. idf.org/aboutdiabetes/what-is-diabetes/facts-figures.html
11 Chen Z, Radjabzadeh D, Chen L, et al. Association of Insulin Resistance and Type 2 Diabetes With Gut Microbial Diversity: A Microbiome-Wide Analysis From Population Studies. *JAMA Network Open* 2021; **4**(7): e2118811-e.
12 Yuan J, Hu Y J, Zheng J, et al. Long-term use of antibiotics and risk of type 2 diabetes in women: a prospective cohort study. *Int J Epidemiol* 2020; **49**(5): 1572–81.

13 Zarrinpar A, Chaix A, Xu Z Z, et al. Antibiotic-induced microbiome depletion alters metabolic homeostasis by affecting gut signaling and colonic metabolism. *Nat Commun* 2018; **9**(1): 2872.
14 Trehan I, Goldbach H S, LaGrone L N, et al. Antibiotics as part of the management of severe acute malnutrition. *N Eng J Med* 2013; **368**(5): 425–35.
15 Ferguson J. Maternal microbial molecules affect offspring health. *Science* 2020; **367**(6481): 978–9.
16 Ohland C L, Pankiv E, Baker G, Madsen K L. Western diet-induced anxiolytic effects in mice are associated with alterations in tryptophan metabolism. *Nutr Neurosci* 2016; **19**(8): 337–45.
17 Demir M, Lang S, Hartmann P, et al. The fecal mycobiome in non-alcoholic fatty liver disease. *J Hepatol* 2022; **76**(4): 788–99.
18 Yuan J, Chen C, Cui J, et al. Fatty Liver Disease Caused by High-Alcohol-Producing Klebsiella pneumoniae. *Cell Metab* 2019; **30**(6): 1172.
19 Koopen A, Witjes J, Wortelboer K, et al. Duodenal Anaerobutyricum soehngenii infusion stimulates GLP-1 production, ameliorates glycaemic control and beneficially shapes the duodenal transcriptome in metabolic syndrome subjects: a randomised double-blind placebo-controlled cross-over study. *Gut* 2022; **71**(8): 1577–87.
20 Mocanu V, Zhang Z, Deehan E C, et al. Fecal microbial transplantation and fiber supplementation in patients with severe obesity and metabolic syndrome: a randomized double-blind, placebo-controlled phase 2 trial. *Nat Med* 2021; **27**(7): 1272–9.
21 Matsumoto M, Inoue R, Tsukahara T, et al. Voluntary running exercise alters microbiota composition and increases n-butyrate concentration in the rat cecum. *Biosci Biotechnol Biochem* 2008; **72**(2): 572–6.
22 McNamara M P, Cadney M D, Castro A A, et al. Oral antibiotics reduce voluntary exercise behavior in athletic mice. *Behav Process* 2022; **199**: 104650.
23 Munukka E, Ahtiainen J P, Puigbó P, et al. Six-Week Endurance Exercise Alters Gut Metagenome That Is not Reflected in Systemic Metabolism in Over-weight Women. *Front Microbiol* 2018; **9**.
24 Grossman J A, Arigo D, Bachman J L. Meaningful weight loss in obese postmenopausal women: a pilot study of high-intensity interval training and wearable technology. *Menopause* 2018; **25**(4): 465–70.
25 Mazier W, Le Corf K, Martinez C, et al. A New Strain of Christensenella minuta as a Potential Biotherapy for Obesity and Associated Metabolic Diseases. *Cells* 2021; **10**(4).
26 Carlsson L M S, Sjoholm K, Jacobson P, et al. Life Expectancy after Bariatric Surgery in the Swedish Obese Subjects Study. *N Engl J Med* 2020; **383**(16): 1535–43.

27 Tao W, Artama M, von Euler-Chelpin M, et al. Colon and rectal cancer risk after bariatric surgery in a multicountry Nordic cohort study. *Int J Cancer* 2020; **147**(3): 728–35.

28 Raglan O, MacIntyre D A, Mitra A, et al. The association between obesity and weight loss after bariatric surgery on the vaginal microbiota. *Microbiome* 2021; **9**(1): 124.

29 West K A, Kanu C, Maric T, et al. Longitudinal metabolic and gut bacterial profiling of pregnant women with previous bariatric surgery. *Gut* 2020; **69**(8): 1452–9.

30 IFS. Living standards, poverty and inequality in the UK: 2021. ifs.org.uk/publications/15512

31 Khan Mirzaei M, Khan M A A, Ghosh P, et al. Bacteriophages Isolated from Stunted Children Can Regulate Gut Bacterial Communities in an Age-Specific Manner. *Cell Host Microbe* 2020; **27**(2): 199–212.e5.

32 Mirzaei M K, Maurice C F. Ménage à trois in the human gut: interactions between host, bacteria and phages. *Nat Rev Microbiol* 2017; **15**(7): 397–408.

33 Vonaesch P, Morien E, Andrianonimiadana L, et al. Stunted childhood growth is associated with decompartmentalization of the gastrointestinal tract and overgrowth of oropharyngeal taxa. *Proc Natl Acad Sci U S A* 2018; **115**(36): E8489–e98.

34 Smith M I, Yatsunenko T, Manary M J, et al. Gut microbiomes of Malawian twin pairs discordant for kwashiorkor. *Science* 2013; **339**(6119): 548–54.

35 Blanton L V, Barratt M J, Charbonneau M R, Ahmed T, Gordon J I. Childhood undernutrition, the gut microbiota, and microbiota-directed therapeutics. *Science* 2016; **352**(6293): 1533.

36 Raman A S, Gehrig J L, Venkatesh S, et al. A sparse covarying unit that describes healthy and impaired human gut microbiota development. *Science* 2019; **365**(6449).

37 Chen R Y, Ahmed T, Gordon J I. A Microbiota-Directed Food Intervention for Undernourished Children. Reply. *N Engl J Med* 2022; **386**(15): 1484.

15. Hacking the Symbiont

1 Haiser H J, Gootenberg D B, Chatman K, Sirasani G, Balskus E P, Turnbaugh P J. Predicting and manipulating cardiac drug inactivation by the human gut bacterium Eggerthella lenta. *Science* 2013; **341**(6143): 295–8.

2 Huang J, Liu D, Wang Y, et al. Ginseng polysaccharides alter the gut microbiota and kynurenine/tryptophan ratio, potentiating the antitumour effect of

antiprogrammed cell death 1/programmed cell death ligand 1 (anti-PD-1/PD-L1) immunotherapy. *Gut* 2022; **71**(4): 734–45.

3 de Groot S, Pijl H, van der Hoeven J J M, Kroep J R. Effects of short-term fasting on cancer treatment. *J Exp Clin Cancer Res* 2019; **38**(1): 209.

4 Vernieri C, Fucà G, Ligorio F, et al. Fasting-Mimicking Diet Is Safe and Reshapes Metabolism and Antitumor Immunity in Patients with Cancer. *Cancer Discov* 2022; **12**(1): 90–107.

5 Ferrere G, Tidjani Alou M, Liu P, et al. Ketogenic diet and ketone bodies enhance the anticancer effects of PD-1 blockade. *JCI Insight* 2021; **6**(2).

6 Haq S, Wang H, Grondin J, et al. Disruption of autophagy by increased 5-HT alters gut microbiota and enhances susceptibility to experimental colitis and Crohn's disease. *Sci Adv* 2021; **7**(45): eabi6442.

7 Vervier K, Moss S, Kumar N, et al. Two microbiota subtypes identified in irritable bowel syndrome with distinct responses to the low FODMAP diet. *Gut* 2021; **71**(9): 1821–30.

8 Hill C, Guarner F, Reid G, et al. The International Scientific Association for Probiotics and Prebiotics consensus statement on the scope and appropriate use of the term probiotic. *Nat Rev Gastroenterol Hepatol* 2014; **11**(8): 506–14.

9 Buziau A M, Soedamah-Muthu S S, Geleijnse J M, Mishra G D. Total Fermented Dairy Food Intake Is Inversely Associated with Cardiovascular Disease Risk in Women. *J Nutr* 2019; **149**(10): 1797–804.

10 Fortune Business Insights. Kefir market size, share and COVID-19 impact analysis. 2020. www.fortunebusinessinsights.com/kefir-market-102463

11 Canale F P, Basso C, Antonini G, et al. Metabolic modulation of tumours with engineered bacteria for immunotherapy. *Nature* 2021; **598**(7882): 662–6.

12 Song L, Xie W, Liu Z, et al. Oral delivery of a Lactococcus lactis strain secreting bovine lactoferricin-lactoferrampin alleviates the development of acute colitis in mice. *Appl Microbiol Biotechnol* 2019; **103**(15): 6169–86.

13 Cubillos-Ruiz A, Alcantar M A, Donghia N M, Cardenas P, Avila-Pacheco J, Collins J J. An engineered live biotherapeutic for the prevention of antibiotic-induced dysbiosis. *Nat Biomed Eng* 2022; **6**(7): 910–21.

14 Hwang I Y, Koh E, Wong A, et al. Engineered probiotic Escherichia coli can eliminate and prevent Pseudomonas aeruginosa gut infection in animal models. *Nat Commun* 2017; **8**(1): 15028.

15 Salminen S, Collado M C, Endo A, et al. The International Scientific Association of Probiotics and Prebiotics (ISAPP) consensus statement on the definition and scope of postbiotics. *Nat Rev Gastroenterol Hepatol* 2021; **18**(9): 649–67.

16 Scheiman J, Luber J M, Chavkin T A, et al. Meta-omics analysis of elite athletes identifies a performance-enhancing microbe that functions via lactate metabolism. *Nat Med* 2019; **25**(7): 1104–9.

17 Zhao X, Zhang Z, Hu B, Huang W, Yuan C, Zou L. Response of Gut Microbiota to Metabolite Changes Induced by Endurance Exercise. *Front Microbiol* 2018; **9**: 765.

16. The Trillion-Dollar Product

1 Cipherbio. 64 Microbiome Biotechs Raise $1.6 Billion in Investments. 2021. blog.cipherbio.com/microbiome/
2 Wang T, Narayanaswamy R, Ren H, Gillespie J W, Petrenko V A, Torchilin V P. Phage-derived protein-mediated targeted chemotherapy of pancreatic cancer. *J Drug Target* 2018; **26**(5-6): 505–15.
3 Gibson D G, Glass J I, Lartigue C, et al. Creation of a bacterial cell controlled by a chemically synthesized genome. *Science* 2010; **329**(5987): 52–6.
4 Hutchison C A III, Chuang R Y, Noskov V N, et al. Design and synthesis of a minimal bacterial genome. *Science* 2016; **351**(6280): aad6253.
5 Wang T, Narayanaswamy R, Ren H, Gillespie J W, Petrenko V A, Torchilin V P. Phage-derived protein-mediated targeted chemotherapy of pancreatic cancer. *J Drug Target* 2018; **26**(5–6): 505–15.
6 Lewin H A. Memories of Carl from an improbable friend. *RNA Biol* 2014; **11**(3): 273–8.

Glossary

Abiosis: The absence of life

Aerobic: Can survive and grow in oxygen

Amensalism: An association between organisms of two different species in which one is inhibited or destroyed and the other is unaffected

Anaerobic: Can only survive in an environment without oxygen

Apoptosis: The process of programmed cell death

Archaea: An ancient domain of single-celled organisms; these microorganisms lack cell nuclei and are therefore prokaryotes

Bioavailability: Refers to the fraction of an administered active drug or metabolite that reaches the systemic circulation; an intravenous medication has a 100 per cent bioavailability

Biofilm: A consortium of microorganisms in which cells stick to each other and often also to a surface; these adherent cells become embedded within a slimy extracellular matrix

Clade: A group of organisms composed of a common ancestor and all its lineal descendants

Commensalism: A long-term biological interaction in which members of one species gain benefits while those of the other species neither benefit nor are harmed

Comorbidity: The presence of two or more medical conditions in a person or patient

DNA: Deoxyribonucleic acid is a self-replicating material that is present in nearly all living organisms as the main constituent of chromosomes; it is the carrier of genetic information

Epigenetics: How cells control gene activity without permanently mutating or changing the DNA sequence itself

Eukaryote: An organism with complex cells, or a single cell with complex structures that exist within it

Exposome: The measure of all the exposures of an individual in a lifetime and how these relate to health

Family: A taxonomic group of one or more genera, especially sharing a common attribute, e.g. Ruminococcaceae

Flagellum: A hairlike appendage that protrudes from microorganisms to provide motility

FMT: Faecal microbiota transplant – the administration of a solution of faecal matter from a donor into the intestinal tract of a recipient

Fungus: Any member of the group of eukaryotic organisms that includes microorganisms such as yeasts and moulds, as well as the more familiar mushrooms

GEM line: A theoretical description of genome–exposome–microbiome interactions during the course of an individual's lifetime or across generations

Genome: The entire set of DNA instructions found in a cell

Genomics: An interdisciplinary field of biology focusing on the structure, function, evolution, mapping and editing of genomes

Genus (pl. genera): A principal taxonomic category that ranks above species and below family and is denoted by a capitalized Latin name in italics, e.g. *Lactobacillus*

Helminth: An invertebrate characterized by an elongated, flat or round body, such as a tapeworm, which acts as a parasite – the stuff of nightmares!

Holobiont: An assemblage of a host and the many other species living in or around it, which together form a discrete ecological unit

Lipopolysaccharide: A large molecule made of lipid and polysaccharide components that acts as a bacterial toxin

Metabonome: A complete set of metabolically regulated elements in a cell, tissue or organism

Metagenome: A collection of genomes and genes from the microbiota

Metagenomics: The study of the structure and function of entire nucleotide sequences isolated and analysed from all the organisms in an organ, community or environment

Metataxonome: A 16S rRNA gene inventory, used to define the microbiota

Microbe: A microorganism or very small living thing that can only be seen with a microscope

Microbiome: The characterization of an entire habitat including all microbes, their genomes and surrounding environmental conditions

Microbiota: Qualitative and quantitative analysis of microorganisms present within a defined environment

Mycelium: The root-like structure of a fungus, consisting of a mass of branching, threadlike hyphae

Nematode: Roundworms constitute the phylum Nematoda and typically occur as parasites in animals and plants, or as free-living forms in soil, fresh water or marine environments

Neurotransmitter: Chemical messengers that carry signals between nerve cells, or between nerve cells and a target organ (e.g. a muscle or a gland)

Pathogen: A microbe or 'germ' that has the potential to cause a disease

Phage: A virus that solely kills and selectively targets bacteria

Phenotype: An individual's observable traits

Phylogenomics: Analysis that involves genome data and evolutionary reconstructions

Phylosymbiosis: The degree of similarity between species microbiomes that recapitulates to a significant extent with their evolutionary history

Phylum (pl. phyla): A level of classification or taxonomic rank below kingdom and above class. In 2021 the International Committee on Systematics of Prokaryotes (ISCP) voted to include the rank of phylum under taxon. Officially, Firmicutes are now Bacillota; Proteobacteria are Pseudomonadota; Actinobacteria are Actinomycetota; and Bacteroidetes are Bacteroidota. However, given that all of the studies I have discussed used the phylum rank, I have used this descriptive term throughout.

Postbiotic: A preparation of inanimate microorganisms and/or their components that confers a health benefit on the host

Prebiotic: A substrate that is selectively utilized by host microorganisms, conferring a health benefit

Probiotic: Live microorganisms that, when administered in adequate amounts, confer a health benefit on the host

Prokaryote: A microscopic single-celled organism that has neither a distinct nucleus with a membrane nor other specialized organelles

Proteome: The entire complement of proteins that is, or can be, expressed by a cell, tissue or organism

Quorum sensing: The regulation of gene expression response in bacteria to fluctuations in cell-population density. This is performed through a form of chemical signalling

RNA: Ribonucleic acid is a polymeric molecule that is essential in various biological roles in coding, decoding, the regulation and expression of genes

Species: Within microbiology, this is surprisingly difficult to define! It is generally accepted as a monophyletic group of bacteria isolates with genomes that exhibit at least 95 per cent pair-wise average-nucleotide-identity (ANI)

Symbiont: An organism living in symbiosis with another

Synbiotic: A mixture comprising live microorganisms and substrate(s) selectively utilized by host microorganisms, which confers a health benefit on the host

Taxon (pl. taxa): A taxonomic group of any rank, such as a species, family or class

Transcriptome: The full range of messenger RNA, or mRNA, molecules expressed by an organism

Virus: A submicroscopic infectious agent that replicates only inside the living cells of an organism

Xenometabolite: A chemical substance found within an organism that is not naturally produced, or expected to be present, within that organism

Zoonosis: An infectious disease that is transmitted between species from animals to humans

Acknowledgements

First and foremost: thanks, Mum and Dad, for the bugs and for not giving up on me when chemistry did. As I have now explained, we can safely put my A-level performance down to my microbiome.

James Harding, mercifully, did not have to undergo an FMT in preparation for this book, but it would not have happened without him and the team at Tortoise Media. You all have my everlasting gratitude. Many of the ideas and concepts in this book have come from discussions with my outstanding colleagues at Imperial College London. I would particularly like to thank Jeremy Nicholson, Elaine Holmes, Sanjay Purkayastha, Julian Marchesi and Ben Mullish for their wisdom and good humour, and all of my past supervisors, who have infused me with passion for this mind-bending topic. I would also like to acknowledge all of those PhD students whom I have been lucky enough to work with over the years. Thanks also goes to Steve O'Keefe, Wouter de Jong, John Alverdy and my other international collaborators, who have minds far sharper than mine. You are a constant source of inspiration.

I must also thank those people who have entrusted their precious tissue to my research over the years, and the patients that I am still lucky enough to serve at Imperial College NHS Trust. I am deeply grateful to both Heather and Ray, who generously allowed me to tell their stories in this book. Ray sadly passed away in 2021, and this book is a tribute to him.

Finally, if you haven't worked it out already, this book is for my children Jacomo and Liberty. I hope this explains things. And it simply couldn't have happened without Sophie, to whom I owe everything. You are my world.

Index

Figures in italics refer to illustrations.

Abbott Laboratories 101
AbbVie 191
abiogenesis 19–20
abiosis 189–90, 275, 309
acidity (pH) 47, 76, 132, 197, 203, 221, 262, 267
Acinetobacter baumannii 164, 179
Acne vulgaris 74
Actinobacteria 12, 74, 131, 132, 165, 177, 183, 311
Actinomycetales 177
addiction 113–14, 125, 195–6, 229–30
additives, food 225–6
adrenaline 36, 59
Adventists 70
Afghanistan 164–5, 168
African Americans 98, 213, 221
agarose 210
ageing 26, 111, 124–42, 203, 229; Alzheimer's/dementia, increasing incidence of xi, 80, 137–9; blue zones 134–5; comorbidities and 126, 134; death 140–42; defined 126; fertility and 86; fountain of youth, search for 124–5; GEM line 126–31, *127*; gut–lung axis and 131–3; inflammation and 129, 132, 134, 135, 136–7, 140, 142; longevity, rules for 125–6; meta hypothesis 139–40; microbiome, ageing 133–6; mosaic of 136
Agent Orange 168
agribiome 148–9, 273 *see also* farming
air pollution 153, 268
Akkermansia muciniphila 138, 215, 223, 238, 247, 258, 263, 265–6
Alcanivorax 147
alcohol 79, 122, 135, 195–6, 228–30, 244–5; alcohol fermentation syndrome 230; alcoholic hepatitis 229

alginate oligosaccharide (AOS) 88
allergic rhinitis (hay fever) 54–5, 66
allergies xi, 48, 53, 54–5, 58, 59, 61, 62–4, 66–7, 69–70, 98, 139, 140, 152, 155, 182, 184, 186–7, 193, 225, 240 *see also individual allergy name*
Alliance of World Scientists 146
alopecia universalis 196, 198
Altos 124
alveoli 131
Alzheimer's disease xi, xx, 80, 81, 137–8, 151
amensalism 48, 53, 57, 59, 128, 155, 165, 254, 275, 309
American Medical Association 199
amino acids 20, 34, 84, 94, 112, 213, 227, 243, 256
Amish 63–4
amoxicillin 171
amputation 161
amylase 31
amyloid plaques 137–8
amyotrophic lateral sclerosis (ALS) 138–9
anaerobic life 9, 21, 67, 98, 163, 266, 309
Anaerobutyricum hallii 263
Anaerobutyricum soehngenii 245–6
anaphylaxis 55
androgens 82, 83, 84
animal fats 214–16, 245, 267
Anisakis nematode 210
anoa 29, 35
Anthropocene 26–7, 52–3
antibiotics: addiction and 114, 195; allergy and 48, 63, 64, 66, 70; ALS and 138; antimicrobial paradox 181–5; bans on 179; Body Mass Index and 81; brain and 80,

315

antibiotics – *cont.*
81, 122; chemotherapy and 196, 197, 198; chronic disease risk and 48, 181–5, 275; *Clostridium difficile* infection and xvii–xviii; diet/food and 212, 226, 241–2, 256; early life, effects of taking in 81, 104, 174–5, 181–2, 183–5, 189; exercise behaviour and 247; farming and 169–71, 179, 210, 241–2; *H. pylori* and 47, 48, 203; immune system and 59; maternal microbiome and 253, 254; natural 66, 175–6, 208, 226; origins/discovery xvii–xviii, 174–8; prescribing rates x, 175, 179–80, 195; probiotics and 259, 264, 267; resistance xviii, 97, 131, 138, 140, 150, 162, 164–6, 168, 171, 175, 178–81, 187, 198, 210, 212, 272, 276; Second World War and advancement of 161, 162, 163, 164, 165, 167, 168, 170–71, 172–3, 177; sex and 72, 76, 80, 81; stroke and 111; taste and 208; ulcers and 47; unnecessary 48, 267, 276; vaccines and 187–8, 189; weight and 249
antibodies 14, 23–4, 32, 58, 66, 68, 103, 187–8, 191
antigens 23, 28, 57, 58, 66, 128, 136, 188, 200
antimicrobial growth promoters (AGP) 179
antimicrobial peptides (AMPs) 66–7
antioxidants 216, 223, 229
antiseptics 91, 122, 161
anxiety 83, 100, 114, 115–16, 119, 121, 243
appendix ix, x, xii, 30, 275
archaea 9, 10, 16, 17, 24, 35, 96, 146, 221, 270
arsphenamine 176
Artemisia 176
Aslam, Ali Ahmed 210
Aspergillus 223
Aspergillus flavus 14
Aspergillus parasiticus 14
asthma xi, 48, 53, 55, 62, 63, 64, 66, 67, 98, 99, 180, 182, 184, 185, 193
α-Synuclein 137, 138
atherosclerosis 52, 136, 215
atopic dermatitis 66, 184

Attention Deficit Hyperactivity Disorder (ADHD) 115
aureomycin 177
Autism Spectrum Disorders (ASD) xi, 79, 115, 116–18, 119
autoimmune disease xi, xx, 38, 53, 56, 67, 79–80, 139, 140, 181, 184, 191, 239, 241
autophagy 141, 258–9
Avoparcin 179
azithromycin 183, 184
Aztec empire 42, 167

BaAka pygmy rainforest hunter-gatherers 33–4
Bacillus subtilis 20, 136
Bäckhed, Fredrik 236–7
bacteria *see individual species name*
bacteriophage (phage) 15–16, 17, 35, 52, 60, 76, 124, 162–4, 183, 218, 230, 251, 272, 311
Bacteroidaceae 152, 165
Bacteroides caecimuris 153
Bacteroides dorei 240
Bacteroides fragilis 66, 119, 218
Bacteroides intestinalis 226
Bacteroides ovatus 114, 120–21, 263
Bacteroides plebeius 210
Bacteroides thetaiotaomicron 28, 203, 218, 239
Bacteroides uniformis CECT 7771 119
Bacteroides vulgatus 201
Bacteroidetes 12, 30, 92, 132, 133, 137, 165, 168, 183, 218, 226–7, 237, 249, 311
Bancel, Stéphane 273
Bantu 34
bariatric surgery xi, 235–6, 248–50, 254
Barnesiella 229
baseline metabolic rate 246–7
β-cells 239, 241
benthic mats 21
Beria, Lavrentiy 163
Berlin 54–5, 62, 64–5, 67, 68, 140, 152–3, 170; Berlin Wall 54–5, 68
β-glucuronidase 85, 197
Bifidobacterium 94, 102, 137–8, 142, 165, 183, 195, 201, 215, 259, 260

bile xix–xx, 82, 96, 135, 188, 202, 207, 214, 216, 220, 222, 227–8, 238, 242, 243, 244, 248, 249
Bimini 124
biodiversity 21, 36, 37, 62, 84, 92, 145–6, 148, 189–90, 254, 266–7, 275; hypotheses 62
biofertilizer 273
biofilm 31, 36, 78, 131, 147, 208, 218, 271, 309
biofuel 272
bioinformatics xxii, 12, 51
biological warfare 160–67
bioremediation xx
biosecurity 171–3
birth 91–106; first 100 days after 104–6, 139; foetal microbiome 95–7; human microbiome, birth of 91–2, 97–101; maternal microbiome 92–5 *see also* maternal microbiome; milk 101–4
birthing routes 63
Blackadder Goes Forth xv
Black Death 41–2, 44
β-lactam 178
Blaser, Martin 62–3, 64
blood-sugar 95, 206, 218, 240, 241
blue zones 134–5
Body Mass Index (BMI) 81, 229, 235, 236, 241, 267
books, microbes on 3–4
Boston Marathon (2015) 266
bowel cancer xi, xv, 49, 50, 132, 140, 198, 214, 221, 249
brain 107–23; ageing and xi, 137–9, 141; antibiotic use and 182, 193, 195; blood–brain barrier 108, 111, 112, 137, 193; development xvi, 37–9; diet and 29, 205, 208, 214, 219, 223, 224, 238–9, 242–3, 245; digital addictions and 122–3; enteric nervous system (ENS) and 109–10; exercise and 247, 265; gut-brain axis 83, 109–10, 112–13, 114, 116, 123, 141, 165, 166, 205, 219, 238, 242, 245, 249, 265, 268; host-manipulation and 121–2; interactome 107; microglial cells and 80–81; neurocysticercosis 108; neurodiversity and 114–19; neurone mapping 107; neurotransmitters 37, 109,
112, 116, 117, 119, 121, 238, 242, 258, 259, 311; peripheral nervous system and 109; psychobiome 113–14, 122–3; psychobiotics 119–20, 195, 264; size, evolution and 37; sleep cycle and 120–21; stroke and 110–12; synaptic pruning 115
branched-chain amino acids (BCAAs) 213, 238
breast cancer 49, 83–4, 102, 181, 197, 209, 257
breastfeeding 93, 101–4, 105–6, 182, 267
British East India Company 160
British Medical Journal 184–5
British Virgin Islands 145, 149, 157
bronchi 131
bronchiolitis 174
Buddhism 70
Burkitt, Denis 219, 221, 253
Byron, Lord 39

Caenorhabditis elegans 135–6
Caesarean section 63, 98, 100, 105–6, 174
Calment, Jeanne 133
Cambrian era 23
cancer *see individual type of cancer*
Candida albicans 14, 76, 78, 99, 230, 244
Candida krusei 230
carbohydrate 20, 34, 74, 216–19, 220, 222, 226, 228, 229, 231, 242, 245, 259
carbolic acid 161, 169
cardiovascular disease xi, 61, 83, 99, 132, 135, 136, 184–5, 211, 214, 220, 231, 236, 238, 243, 248, 254, 265
carotenoid 222
Ceftriaxone 175
Centers for Disease Control and Prevention (CDC) 183, 235
cereals 32, 93, 216–17, 220, 222
Ceres 11
Chain, Ernst 174
Chambers, Major Peter 164
checkpoint inhibitors 200, 201, 257, 258
chemotherapy 176, 194, 196–8, 257, 272
'chickenization' of Western diet 169
chlamydia 77, 86
Chlamydia trachomatis 85

317

Chlorpyrifos 149
cholera 44, 157, 163
cholesterol 191, 201–2, 203, 214, 215, 216, 220, 243, 257, 262
choline 112, 214, 244
Christensenellaceae 215, 247–8
chronic obstructive pulmonary disease (COPD) 45, 131–2
chronic rhinosinusitis (CRS) 66
cinnamon oil 256–7
circadian rhythms 120–21
Claptrap, Operation 164
Cleave, Captain T L: *The Saccharine Disease* 217
climate change 9, 21, 25–6, 38, 53, 64, 66, 68, 145–6, 148, 149–50, 156, 157, 168, 254, 273, 276
Clostridiaceae 34
Clostridium cluster IV 259
Clostridium cluster XI 197
Clostridium difficile (C. diff; renamed *Clostridioides*) ix, xvii–xxi, 183, 203
Clostridium perfringens 28
Clostridium scindens 135
Clostridium sporogenes 121
cloudinomorphs 22–3
clownfish 28
Clustered Regularly Interspaced Short Palindromic Repeats (CRISPR) 16, 172
coeliac disease 32, 48
Cold War (1946–91) 69, 164, 170
Coley, William 198–9
Collinsella aerofaciens 211
colon xv, xviii, xix, xxii, 5, 32, 49, 100, 105, 125, 181, 212, 216, 218, 220, 223, 227, 229, 237, 263; cancer 151, 156, 213, 222, 249
colonialism 160
colonoscopy xix, 51
colorectal cancer ix, 49, 151, 220, 221, 257
Columbia 22
Columbian exchange 159
Columbus, Christopher 159, 175
commensalistic relationships 48
comorbidities 61, 126, 134, 309
Compactin 202

conflict 159–73; antibiotics in farming linked to 168–71; antiseptic innovation and 161–6; biological warfare 160, 166–7; biosecurity / COVID-19 171–3; climate change and 166–8; Columbian exchange 159–60; conflictome 168, 173; spice trade and 160
consumption, addiction to 148
contraception 54, 80, 84, 85
cooking 30, 226–8, 232
Cooper, Alan 31
Coprococcus 113
coprolites 32–3
coprophagy 34, 104, 200
coprostanol 216
Corynebacteriaceae 75
Corynebacterium 75
co-speciation 28–9, 34
COVID-19 xii–xiii, 42, 43, 61, 77, 101, 118, 123, 141, 155, 156, 161, 166, 168, 171–2, 186–7, 188, 189, 191, 208–9, 211, 223, 234, 235, 236, 250, 272
Crichton, Michael: *Jurassic Park* 25
Crohn's disease 132, 227
cross-feeding 14, 48, 221
Cullen, William 204
Cutibacterium acnes 74
cyanobacteria 21, 120
cyclooxygenase (COX) 193
cytochrome P450 (CYP450) 193–4, 256
cytolysin 230
cytoplasmic incompatibility 80

da Gama, Vasco 160
dark matter, biological 6–7, 15, 60, 97, 207, 241, 251, 276–7
Darwin, Charles 7–8, 11, 30
Dashiell, Bessie 198–9
dating apps 73, 78
death 140–42
Deepwater Horizon oil spill 146–7
Deinococcus radiodurans 20
Delbrück, Max 163
dementia xi, 46, 48, 137–8, 203
dendritic cells 66, 67

depression xx, 81, 85, 113, 114, 122, 154
d'Hérelle, Félix 162–3
diabetes xi, xx, 86, 95, 149, 220, 231, 236, 238, 243, 248; gestational 95, 249–50; type I 88, 181, 185, 239–41; type II 86, 99, 136, 153, 182, 217, 218, 236, 241–3, 244, 257, 262
Dialister 113
diarrhoea xvii, xix, 44, 103, 113, 116, 165, 187, 196, 197, 245, 251, 259
diet 206–32; additives 225–6; alcohol 228–30 *see also* alcohol; animal fats 214–16, 245, 267; antibiotics and 212, 226, 241–2, 256; brain and 29, 205, 208, 214, 219, 223, 224, 238–9, 242–3, 245; carbohydrate 216–19 *see also* carbohydrate; 'chickenization' of Western diet 169; cooking 226–8; diet–microbiome–drug interactions 255–9; farming practices and *see* farming; fermented foods 223–5; fibre 219–22 *see also* fibre; FODMAP diet 259; global 209–11; gut microbiome evolution and 29–34; junk food 211–15; keto diet 245, 258; liquid diet 204, 210, 227–8; maternal microbiome and 93, 94–5; Mediterranean diet 211; obesity and 236, 237–8, 241, 242, 244, 245, 246; palaeo diet 29, 38; polyphenols 222–3 *see also* polyphenols; probiotics/prebiotics/postbiotics *see* probiotics, prebiotics, postbiotics; protein 212–14 *see also* protein; raw foods 125, 210, 226; restaurant hypothesis 207; sex and 72, 86, 88; sleep and 120–21; soft drinks 230–31; stroke and 111–12; sushi and 210; taste and 208–9; vegetarian 70, 207, 225; Western x, 9, 30, 32, 86, 169, 194, 211, 215, 244, 254
digital addiction 122–3
digital microbiome 269–71
digoxin 255–6
dinosaurs 24, 25–6, 156
disgust xv, 34, 73
dispersal 99
displaced people 160–61, 166, 168
diversification 104–5

Djerassi, Dr Carl 84
DNA 3, 8, 9–10, 11, 16, 20, 21–2, 24, 28, 31, 76, 77, 93, 127, 155, 163, 196, 270–71, 272, 309, 310
dopamine 113–14, 119, 122
Dorea formicigenerans 120
Dorea longicatena 120
drifts 104
driver organisms 49–50
drugs 191–205; chemotherapy 196–8; FMT and 195–6, 197; immunotherapy 198–201; microcebo effect 204–5; number of drugs prescribed globally every year 191; painkillers 192–6; personalized medicine and 192; pharmacokinetics 194; polypharmacy 201–4; substance addiction 194–6 *see also individual drug name*
Duggar, Benjamin 177
duodenum xxi, 246
dust 45, 64, 67, 97, 152
dyslexia 114–15

East Germany 54–5, 62, 67, 140, 170
E. coli 35, 85, 138, 179, 180, 240, 264, 270–72 *see also Escherichia coli*
eczema 62, 180, 184
Ediacaran period 22
Eggerthella lenta 245, 256
Ehrlich, Paul 176–7
eicosanoid 215
Eiseman, Dr Ben xix
Eisenbergiella massiliensis 258
Eliava, Giorgi 163
emulsifiers 225
Endo, Akira 201–2
endocrine system 112, 147
endometriosis 83
endosymbiosis 22, 23
engineering, microbiome 268, 269–73
enteric nervous system (ENS) 109–10, 112, 259
Enterobacteriaceae 117, 152, 197, 251
Enterococcus faecalis 230
Enterococcus faecium 179
enterovirus B 241
environmental exposures *see* exposome

eosinophilic oesophagitis 66
epigenetics 120, 127, 129, 139, 309
Escherichia coli 9, 14, 103, 163, 240, 263
    see also E. coli
Estonia 240
Eubacterium 111
Eubacterium ramulus 84
Eubacterium rectale 150
eukaryotes 8, 10, 11, 13, 14, 21, 24, 272, 309
European Union (EU) 55, 132, 179
exercise/physical activity 246–8, 255, 265–6, 267–8, 275
exposome 27, 30, 33, 38, 64, 68–9, 70, 79, 116, 122–3, 126, 127, 128, 129, 130–31, 140, 143–232, 253, 254, 267, 268, 269, 270, 309; antibiotics *see* antibiotics; diet *see* diet; drugs *see* drugs; global microbiome *see* global microbiome; military conflict *see* military conflict; vaccination *see* vaccination
extinction rate, microbiome 33
extremophiles 9, 19, 96

Faecalibacterium 113, 202, 265
Faecalibacterium prausnitzii 120, 150, 211, 213, 259, 263
faecal microbiota transplant (FMT) xv–xxi, 59–60, 70, 232, 251, 252, 273, 310; addiction and 195–6; chemotherapy-induced peripheral neuropathy (CIPN) and 197; dementia and 138; fertility and 88; metabolic syndrome and 245, 246; pneumonia and xviii–xxi, 275; precision therapy 268; vaccines and 187
Falkow, Stanley 62–3, 64
family microbiome networks 153–6
farming 31–2, 39, 41, 64, 148–9, 158, 167, 207; antibiotics in 169–71, 172, 179, 181, 189
farting 33, 221
fermented foods xv, xxii, 45, 134–5, 221, 223–5, 227, 228–9, 230, 246, 249, 250, 259, 261, 262, 267
fertility 21, 61, 75, 83, 86–8, 89, 147–8
fibre 32, 33, 70, 94, 111, 120, 159, 188, 207, 212, 219–22, 223, 224, 225, 227, 228, 235, 242, 245, 246, 247, 264–5, 267

fight or flight response 36, 109
Fiji Community Microbiome Project (FijiCOMP) 154–5
Finland 81, 102, 163, 185, 240
Firmicutes 12, 34, 92, 94, 131, 132, 133, 137, 168, 237, 248–9, 311
Fleming, Sir Alexander xvii, 174, 175, 177
Florey, Howard 174
Fly Formula, Operation 101
FODMAP diet 259
foetal microbiome 95–7
folate 93–4, 257
food *see* diet
food allergy 55, 63, 69–70, 184, 225
Food Allergy Program, Boston Children's Hospital 69–70
Fox, George 9
free radicals 111
'friendly' microbes 47
fructose 211, 218, 222, 231, 239, 244, 259
fruit fly (*Drosophila melanogaster*) 72
fungi 8; alcohol and 229; circadian rhythms and 120; evolution of 22; fermented foods and 223, 224; gut and (mycobiome) 13–14; NAFLD and 244 *see also individual species name*
Fusobacterium nucleatum 32, 49, 198, 209

Gardnerella 76
gastric cancer 47, 203
gastric stomach xxii, 203, 206, 235
gastrointestinal infections 43–4, 113, 116, 184
gender/genderome 74, 79–82, 84–5, 88–90, 182, 248 *see also* sex
gene editing 16
Generation Z xi
Genghis Khan 166
genital microbiota 75–6, 88
genitourinary syndrome of menopause (GSM) 85
genome–exposome–microbiome (GEM) interactions 69, 310; GEM disease 253–4, 271; GEM line 126–31, 127, 136, 142, 152, 173, 268, 310

germ-free animals 80, 100–101, 105, 115, 117, 138, 195, 197, 215, 230, 237, 244, 251–2, 256, 269
germ theory xiii, 45, 53, 161
Giardia duodenalis 13
ginseng 257
globalization x, xii, 38, 55, 68, 130, 153–4, 157, 167, 211, 236, 276
global microbiome 145–58, 160, 209–11, 266; biodiversity loss 148–9; climate change 145–6, 148, 149–50, 156, 157 *see also* climate change; consumption, addiction to 148; ethnic diversity and 149–51; family microbiome networks 153–6; fourth age of man and 145–9, 157–8; industrial spills 146–7; plastic waste 147–8, 153, 157 *see also* plastic; urbanization 40, 151–3
GLP-1 238, 246
gluten 32
glycaemic index 217–18
gonadotropin-releasing hormone (GnRH) 87
gonorrhoea 77, 86, 164, 177, 179
Gordon, Jeff 236–7, 251, 252
Gramicidin 170
Greater Baltimore 149
Great Ormond Street Hospital xvii, 184–5
Gryllus texensis 78
Gulf War syndrome 165–6, 173
gut-associated lymphoid tissue (GALT) 128
gut microbiome xvi; ageing and *see* ageing; antibiotics and *see* antibiotics; birth and 91–106; brain and *see* brain; conflict and *see* conflict; dark matter within 6–7, 13–14, 15, 60, 97, 207, 241, 251, 276–7; defined xvi; diet and *see* diet; drugs and *see individual drug name*; engineering 268, 269–73; evolution of 22–3, 29–34; exposome and *see* exposome; faecal microbiota transplant (FMT) and *see* faecal microbiota transplant (FMT); global microbiome and *see* global microbiome; gut resistome 181; inflammation and *see* inflammation;
leaky gut syndrome 65–8; obesity and *see* obesity; physical distribution of xxi–xxii; sex and 71–90; synthetic biology and 172, 271–3; training 265–6

haem xv
Haloferax massiliense 9
Hamin Mangha, China 39, 40, 41
Handelsman, Jo 11
hand sanitizer 4
Hata, Sahachiro 176
heart disease 45, 120, 152, 214
Heidelberg University 203
Helicobacter pylori 28, 47, 48, 179, 203, 210
helminths 13, 17, 52, 60, 61, 159, 251, 310
Herodotus 124
high-density lipoproteins (HDL) 215
hippocampus 242
HMG-CoA reductase 202
holobiont 27–9, 38, 89, 90, 156, 157, 218, 269, 276, 310
holosexual microbiome 88–90
Homo sapiens, emergence of 26, 29, 30–32
honey 176
hookworm 60
horizontal gene transfer 24, 29, 34, 168, 178
hormone replacement therapy (HRT) 85
hospital-acquired infection xvii, 179
host-manipulation 121–2
HSV-1 78
Human Genome Project (HGP) 10, 11
Human Microbiome Project 12, 81
human milk oligosaccharides (HMOs) 103
human papillomavirus (HPV) 49–50, 59, 77, 78, 186
Humira 191
hunter-gatherers 34, 39
Hutterites 63–4
hydrogen cyanide 36
hydrolase 148, 218
hygiene hypothesis 62–3
hypercholesterolemia 202

Ideonella sakaiensis 148
IgM 24, 58

Ignicoccus hospitalis 9
immigration 156, 209–10
immune system xii, xvi, xxii; *Acne vulgaris*
  and 74; adaptive 23–4, 57–9, 60, 62, 66, 67,
  83, 96, 118, 136, 223, 239; ageing and 128,
  130, 132–3, 135, 136–7, 139–40; antibiotics
  and 188, 189, 191, 194; appendix and 30;
  autoimmune disease *see* autoimmune
  disease; birth and 94, 96, 97, 98, 99, 100,
  103, 104; brain and 109, 110, 112, 118, 121;
  conflict and 160, 165, 166, 170; diet and
  212, 214, 215, 223; environment and 147–8;
  evolution of 23–4; fertility and 88; FMT
  and 59–60; function of 56–7; gluten and
  32; immunotherapy 58, 198–201, 258, 263;
  immune-senescent 136–7;
  inflammasome 57; inflammation and *see*
  inflammation; innate 56–9, 60, 64, 65, 66,
  67, 69, 83, 96, 112, 128, 136, 214, 223, 239;
  intestinal parasites and 60–61; leaky gut
  and 65–8; liver and 243, 244;
  microbiome–immune axis 56–70;
  obesity/weight and 239, 240, 241; STIs
  and 78; 'tone' 58–9, 212; tumour
  microenvironment and 50
Immunoglobulin A (IgA) 58, 66, 83
immunotherapy 58, 198–201, 258, 263
Imperial College London xviii, 194, 249–50
India 160, 176, 180, 191, 209–10
indigenous peoples 33–4, 151, 153, 159–60,
  166–7
indole-3-propionic acid (IPA) 121
industrial spills 146–7
infant formula 101–4, 174
inflammation ix; *Acne vulgaris* and 74; ageing
  and 129, 132–3, 134, 135, 136–7, 140, 142;
  allergy and 54–5, 64, 68–9; antibiotics and
  180, 181, 185, 187, 188–9; blood nematode
  infections and 60; brain and 111–12, 113,
  114, 118, 120; *C. diff* and xvii–xviii, xx;
  chemotherapy and 196, 197; coeliac
  disease and 32; conflict and 165, 166;
  COVID-19 and 61, 209; defined 59; diet
  and 142, 213, 214, 215, 216, 218, 222, 223,
  224, 225, 227, 229, 230, 256, 258–9, 263–4;

immune system and inflammatory
  response 57–60, 68–9; leaky gut and 65–8;
  modern inflammatory pandemic 68–9;
  obesity and 236, 239, 240–41, 242–3, 244,
  245, 247, 249, 253, 254; paracetamol and
  193; pregnancy and 95; vagina and 78, 84
inflammasome 57
inflammatory bowel disease ix, 114, 132,
  155, 181, 191, 203, 225, 227
insulin 85–6, 95, 141, 213, 215, 238, 239–42,
  245–6, 248, 254, 258, 264
interactome 107
intermittent fasting 121, 141–2, 245
International Health Metrics Evaluation 151
International Organization for Migration 156
International Space Station 20
intestine xvii, 9, 23, 30, 61, 82, 92, 95, 96,
  100, 105, 110, 133, 135, 171, 214, 217;
  evolution of 22–3
Iraq War (2003–11) 162, 164–5, 168
irritable bowel syndrome (IBS) ix, xx, 114,
  225, 259, 264
Ivermectin 61
in vitro fertilization (IVF) 87, 270

Jackson Laboratory (JAX) 200–201
Jennings, Margaret 174
Jet Propulsion Laboratory 20
Joint Biosecurity Centre (JBC) 172
junk food 121, 211–14, 239, 243

Kanner, Leo 116
Karelia 240
kefir 223, 262–3
Keller, Josbert xix
keto diet 245, 258
Kinross, Jacomo 91, 98
Kinross, Liberty 174, 175
kissing 3, 31, 71, 73, 76–7, 128, 267
Klebsiella pneumoniae 179, 244–5
Koch, Robert 46–7, 49, 52, 176, 199
Kwashiorkor 251–2

Laboratoire du bactériophage 162–3
Lachnospiraceae 152, 257

Lactobacillus 76, 85, 87–8, 111, 142, 215, 249, 256, 310
Lactobacillus bulgaricus 135, 262, 266
Lactobacillus casei 260
Lactobacillus crispatus 76, 78, 87
Lactobacillus iners 76, 78
Lactobacillus plantarum 72, 215–16
Lactobacillus reuteri 119, 218, 260
La Cueva de los Muertos Chiquitos, Mexico 32–3
L-dopa 203
leaky gut syndrome 65–8
Leang Bulu' Sipong, Sulawesi 29, 35, 36–7
Leeuwenhoek, Antoine van 16–17
León, Juan Ponce de 124
leucine 213
leukaemia xvii, 60
Lewy Body disease (LBD) 137
LGBTQ+ 89
life, emergence of 19–21
lipase 74, 214
lipids 20, 93, 103, 185, 201, 202, 214–16, 243, 245, 264, 267, 310
lipopolysaccharides 57, 208, 240, 310
liquid diet 204, 210, 227–8
Lister, Joseph 161
liver xix, xx, 49, 80, 93, 94, 120, 135, 139, 182, 193–4, 198, 214–15, 230, 237, 258; gut–liver axis 243–6, 249
love 71–5, 90
lung xiii, xvii, 5, 15, 43, 52, 67, 75, 129, 139, 151, 182, 269; cancer 131, 197; gut–lung axis 131–3, 249
lycopene 222
lymphocytes 24, 57, 67, 96, 118, 128

Macau University of Science and Technology 257
macroalgal/microalgal foods 273
macrolide 184–5
macrophages 57–8, 67
malaria 25–6, 41, 42, 156, 176, 188
Malassezia 74
Malassezia furfur 13–14

malnutrition xx, 32, 55, 133, 196, 229, 242, 250–53, 254
mammals, emergence of 26
mammary glands, emergence of 26
Manila, Philippines 151–2
Margulis, Lynn 22, 27
Marinobacter 147
Mars Sample Return Mission 20
Marshall, Barry 47
Massachusetts General Hospital 84
mastadenovirus C 241
maternal microbiome 62–3, 91–106, 118, 123, 139, 184, 253–4; breastfeeding 93, 101–4, 105–6, 182, 267; childbirth and 91–106; environmental influences on 62–3, 94, 99, 105, 106; fertility and 86–8, 89; first 100 days of babies' life and 104–6, 139; foetal microbiome and 95–7; milk 101–4; pregnancy 43, 50, 80, 87–8, 90, 92–5, 96, 98, 118, 193, 242, 249; premature babies 99–100; sterile-womb hypothesis 92
M cells 110
meat 29–30, 31, 169, 170–71, 179, 206, 207, 213–14, 221–2, 223, 232, 267, 273
meconium 96
Mediterranean diet 211
menopause 82, 84–5, 90
menstrual cycle 74, 83, 84–5, 129
metabolites 14, 37, 52–3, 57, 63, 66, 69, 93, 94, 108, 177, 193–4, 197, 212, 240, 242, 244, 250, 259, 263, 264, 266, 270, 309, 312; ageing and 138–9, 140; brain and 112, 114, 117, 119, 120, 121; diet and 215, 221–2, 227–8, 238
metabonomics 12
MetaCardis study 202–3
metagenome/metagenomics xxii, 10–12, 16, 30–31, 33, 148, 152, 178, 180, 217, 263, 310
Metagenomics and Metadesign of Subways and Urban Biomes (MetaSUB) 180
meta hypothesis 139–40, 253
Metchnikoff, Élie 124–5, 134, 210
metformin 257–8
methane 9, 33, 171, 214, 221

323

Methanobrevibacter oralis 31
Methanobrevibacter smithii 33
methanogens 9, 171, 221
Methanopyrus kandleri 9
Methicillin-resistant *Staphylococcus aureus* (MRSA) 179
Methylobacterium ajmalii 20
microbiota-directed complementary food (MDCF) 252, 254, 268
Microbrachius dicki 24, 25
microcebo effect 204–5
microglia 80–81, 112, 197
microplastics 67, 147–8
middle age 19, 81–2, 228
military x, 161–73
milk 26, 54, 101–4, 114, 170, 184, 214, 224–5, 262–3
Miocene period 26
money, emergence of 39–40
monosaccharides 237, 259
mould 31, 177, 201–2, 271, 310
mTOR (mammalian/mechanistic target of rapamycin) 213, 257
Mulago Hospital, Uganda 219–20
Mullish, Dr Ben xviii, xix
multiple sclerosis (MS) 60
Multiple System Atrophy (MSA) 137
mutualistic relationships 5, 37, 48, 173, 207, 221, 238
myalgic encephalomyelitis (ME) 166
Mycobacterium mycoides 271–2
Mycobacterium neoaurum 85
mycotoxins 14

nanoplastics 147
National Institutes of Health (NIH), US 12
Neanderthals 26, 29, 31, 76
necrobiome 40, 127, 140, 142
Neisseria 132
Neisseria gonorrhoeae 86, 179
nematodes 13, 60–61, 73, 80, 135, 159–60, 183, 210, 311
Neolithic 38, 39, 40, 41, 53
neosalvarsan 177
neurocysticercosis 108

neurodiversity 114–18
neurones 37, 107, 111, 115, 138, 219
neuropod cells 112
neurotransmitters 37, 109, 112, 116, 117, 119, 121, 238, 242, 258, 259, 311
New York University 135
nicotinamide 138–9
Nightingale, Florence 161
9/11 ix–x
nitrogen dioxide (NO_2) 153
nitrogen oxides (NOx) 153
nocebo 204, 205
non-alcoholic fatty liver disease (NAFLD) 244–5, 254
nucleic acids 20
nutritional supplements 256–8

obesity xi, xx, 38, 53, 61, 88, 135, 235–54; air pollution and 153; antibiotics and 182, 185, 201; Body Mass Index (BMI) and 81, 229, 235, 236, 241, 267; COVID-19 and 235, 236, 250; crisis 235–6, 248; diet and 213, 215, 218, 235–54 *see also* diet; exercise and 246–8; FMT 245–6; GEM disease 253–4; genetic risk 236, 237–8, 239–40, 251; gut–liver axis and 243–6; inflammation and 236, 239, 240–41, 242–3, 244, 245, 247, 249–50, 253–4; intermittent fasting and 245; maternal 50, 94–5, 98, 102; men and 85–6; meta hypothesis and 140; military conflict and 165, 166; obese microbiome 236–9; pesticides and 149; sedentary microbiome 246–8; starvation/malnutrition and 250–53; surgery xi, 235–6, 248–50, 254; Syndrome X 239–43; thermodynamic theory of 236; women and 50, 79, 83, 84, 85–6, 94–5, 98, 102
obstructive sleep apnoea 131
oestradiol 82, 83, 84
oestrobolome 82–3
oestrogen 82–5, 87
O'Keefe, Professor Stephen 221, 245
'old friends' hypothesis 62
olfactory chemical communication 73

olive oil 216
Omega-3 (linoleic acid) 215
Omega-6 (α-linolenic acid) 215, 231
Ophiocordyceps unilateralis 121–2
opiates 194–5
oral microbiome 30–31, 93, 209
orchestral signalling 117, 139, 173, 184
Ötzi 27–8, 38, 48
Ovid: *Metamorphoses* 139
Oxalobacteraceae 75
OxyContin 195
oxytocin 119
ozone (O_3) 67, 153, 155

palaeo diet 29, 38
Paleohaemoproteus burmanicus 25
pancreatic cancer 257
panda bears 73, 104
Pangaea 25
panspermia 20
Panton-Valentine leukocidin (PVL) 164–5
Parabacteroides distasonis 138, 201
paracetamol 174, 192–4, 205, 230
parasites xix, 10, 13, 14, 15, 16, 25, 41, 48, 61, 72–3, 89, 159–60
parasympathetic nervous system 109
Parkinson's disease (PD) 79, 137, 203
passenger microbes 49–50, 63, 88
Pasteur Institute 163
Pasteur, Louis 44–5, 52, 100, 161, 186, 223
Patancheru, India 180
p-Cresol 117, 194
PD-1 199–200, 201, 257, 258
PD-L1 201, 199–200
pembrolizumab 199–200
penicillin xvii, 14, 162, 164, 170, 174, 175, 177, 182, 184, 185
Penicillium citrinum Pen-51 202
penis 75–6
periods 83
peripheral nervous system 109, 112
Perseverance rover 20
Persian Gulf War (1990–91) 165
personal data 271

personalized medicine 123, 134, 192, 218, 232, 255, 270
pesticides 149, 157, 165
phage *see* bacteriophage
pharmacokinetics 194
pharmacomicrobiomics 270–71
phenome 128–9
photosynthesis 21
phylosymbiosis 27, 29, 311
Physarum polycephalum 271
Pinggu (PG), Beijing 81
placebo effect xx, 195–6, 204–5
Plasmodium falciparum 41
Plasmodium malariae 41
Plasmodium ovale 41
Plasmodium vivax 41
plastic 64, 67, 100, 103, 128, 147–8, 153, 157, 175, 206, 228, 264, 272, 273, 276
plasticene 147
pneumonia xvii, 43, 45, 174, 189, 244
Poinar, Hendrik 25
polycystic ovary syndrome (PCOS) 83, 84
polyethylene terephthalate (PET) 148
polypharmacy 202–4
polyphenols 159–60, 176, 222–3, 224, 229, 256, 264, 267
polysaccharide 57, 88, 208, 220, 225–6, 240, 257, 310
polyunsaturated fatty acids (PUFAs) 215, 264
postbiotics 265, 311
post-traumatic stress disorder (PTSD) 165, 173
postulates 45–6, 49–52
Pratītyasamutpāda 70, 273
prebiotics 255, 264–5, 268, 311
pre-eclampsia 95
pregnancy 43, 50, 80, 87–8, 90, 92–5, 96, 98, 118, 193, 242, 249
premature babies 98, 99–100
Prevotella 82, 85, 131, 132, 150, 152, 257, 265
Prevotella copri 150, 211
Prevotellaceae 34, 247–8
primates 26, 31, 34, 37, 73, 153

probiotics 14, 87, 88, 119–20, 135, 147, 216, 245, 248, 255, 256, 259–62, 263–5, 266, 268, 270, 273; next-generation (NGPs) 255, 263–4, 273
productive exposome 79
progesterone 83, 84, 85, 87
prokaryotes 8, 11, 24, 35, 88–9, 309, 311
Prontosil 162, 177, 192
prostaglandins 193, 205
proteases 212, 263
protein, dietary 29–30, 32, 103, 111–12, 117, 171, 212–14, 221, 223, 235, 245, 249, 250, 251, 256, 257, 273
Proteobacteria 12, 74, 131, 132, 137, 165, 218, 247, 248–9, 251, 311
proteomics 12, 249
proton-pump inhibitors 203
Pseudomonadaceae 75
Pseudomonas aeruginosa 35, 163, 179, 264
Pseudomonas veronii 28
psilocybin 14
psychobiome 113–14, 122
psychobiotics 119–20, 195, 264
pullorum disease 169
pyridostigmine bromide 165

quantum biology 18
quorum sensing 36, 312

Raymond (pneumonia) xvi, xviii, 183, 203, 275
Reavan, Gerald 243
recombination-activating gene 23–4
rectum xxii, 266
Red Queen hypothesis 24–5, 72–3
refugees 160–61, 166, 168
resident microorganisms 4
resistome 178–81, 182, 275
Respiratory Syncytial Virus (RSV) 174
restaurant hypothesis 207
Resveratrol 229
rhythmicity 120
RNA (ribonucleic acid) 8–10, 12, 20, 93, 273, 312
Roseburia 111, 211

Rothia 132
Royal Society 16–17
Ruminococcaceae 152, 226, 310
Ruminococcus bromii 220–21
Ruminococcus gnavus 211, 225
Ruminococcus torques 138, 211
Russia 163–4, 240

Saccharomyces boulardii 14
Saccharopolyspora 3
St Mary's Hospital, Paddington ix, xvii, 174, 189, 217, 275
salmon farming 171
Salmonella 73, 110
Salmonella enterica 113
Salmonella pullorum 169
Salvarsan 176
Schabowski, Günter 54
Schöneck-Kilianstädten 40
Schweinehochhaus 170
Second World War (1939–45) 8, 69, 130, 162, 163, 164, 167, 169, 170, 173, 177, 192, 209
sedentary microbiome 246–8
segmented filamentous bacteria 118, 240
semen 75, 77, 88
Semmelweis, Ignac Philipp 161
sepsis x, 43, 124, 141, 161, 172, 196
serotonin 112, 116, 119, 122, 238, 258
7α-dehydroxylase 135
sex xxii, 14, 31, 75–90; courtship/love and 71–5; evolution of 24–5, 26; fertility 86–8; gender and 79–82; holosexual microbiome 88–90; hormones 82–5; mating and 71–90; sexually transmitted infection (STI) 49–50, 77, 78, 84 *see also* STI name; sexual reproduction 75–9, 109, 121
Shelley, Mary: *Frankenstein* 273
shift work 120
'shit', word xv
short-chain fatty acids 207, 221–2, 223, 238, 240–41, 242, 247, 256, 258, 259, 265, 266
Sichuan University, China 108
sickle cell disease (SCD) 16, 41
sickness behaviour 113
Singer, Judy 114–15

skin 3–4, 5, 9, 17, 22–3, 39, 52, 73–4, 81, 97, 98, 99, 102, 131, 152, 198–9, 264, 269
sleep xvii, 117, 120–21, 129, 131, 267–8, 275
small bowel xxi–xxii, 32, 65, 206, 212, 217, 220, 227, 235, 242, 248
smallpox 42, 78, 160, 167, 172, 186
smoking 67, 125, 128, 131–3, 267
Snow, Dr John 44, 180
social inclusion/socialization 135, 153, 228, 275–6
soft drinks 230–31
soil 11, 12, 97, 128, 132, 142, 146, 148–9, 167–8, 176, 177, 180, 273
Sorcerer II 11
South Pacific Gyre 26
Southwest Finland Birth Cohort 81
Soviet Union 163–4
soy 114, 214, 223, 224–5
sperm 24, 80, 86–8
spices 160, 210, 256–7, 267
Spirochaetaceae 34
spontaneous generation theory 44–5
Stalin, Josef 163
standard model: of physics 6; of medicine 7
Staphylococcus 73, 75
Staphylococcus aureus 43, 163–5, 179
Staphylococcus epidermidis 73
starch 31, 32, 217, 218, 220–31, 226, 227, 264
starvation 35, 168–9, 171, 231, 238–9, 250–53, 257, 258
statins 191, 201–2, 203
stercobilinogen xv
sterile-womb hypothesis 92
Stone Age 39
Strachan, Professor David 61–2
strep throat 162
Streptococcus 31, 75, 85, 111, 131, 132, 208, 229, 260
Streptococcus agalactiae 92
Streptococcus mutans 31
Streptococcus pyogenes 163
Streptococcus thermophilus 26
Streptomyces 176, 177, 208
Streptomyces aureofaciens 169, 177
streptomycin 177, 178
stroke 45, 52, 79, 110–12, 152, 203, 214, 220

Subdoligranulum 33, 47
substance addictions 194–6
Succinivibrionaceae 266
sugar 34, 57, 120–21, 205, 211–12, 217–19, 225, 226, 228, 229, 230, 231, 241, 242, 244, 252, 257–8, 267
Sulfaquinoxaline 169
sulphides 212, 213
suprachiasmatic nucleus 120
supraorganism trauma care 172–3
surveillance capitalism 271
sushi 210
Sutherland, Alex: *Attempts to Revive Antient Medical Doctrines* 204
symbiosis: ageing and 124, 128; birth and symbiotic bacteria 91, 95, 103, 104; brain and 107–8, 118, 122; co-evolution of symbiotic relationships 28–9; co-speciation 28–9, 34; conflict and 173; cross-feeding 14, 48, 221; diet and 211, 215, 223, 226, 227–8, 239; endosymbiosis 22, 23; hacking the symbiont 255–68; immune system and 58–9, 61; leaky gut and 65, 66, 67; pandemics and 53; phylosymbiosis 27, 29, 311; resistome and 178, 181, 182, 187, 190; sexual microbiome and 75, 76, 77, 78–9; symbiont defined 48; symbiotic bacteria defined 5; wheat and 32
sympathetic nervous system 109, 242
Syndrome X 239–43
synthetic biology 172, 271–3
syphilis 78, 91, 164, 176–7

Taconic Farms (TAC) 200–201
Taenia solium 108
Tanpopo mission 20
taste 208–9
Tenericutes 215
testosterone 82, 84, 85–6
tetracycline 170, 177, 178
TGF-β1 263
3-hydroxybutyrate (3HB) 258
33-mer gliadin 32
tomato 159, 209, 222
Tortola, British Virgin Islands 145, 157–8

trading networks 39–40, 160
Traditional Chinese Medicine (TCM) 176
transbacteria 80
transgender people 80
Treponema pallidum 78
Triassic period 25, 26
Triticum dicoccum 32
Triticum monococcum 32
trimethylamine (TMA) 214, 227
Trimethylamine *N*-oxide (TMAO) 112, 214
Trypillia 40
tryptophan 112, 117, 242, 256
tuberculosis 40–41, 42–3, 46, 177, 179, 186, 188
Tufts University, Massachusetts 231
tumour microenvironment 50, 200, 201, 257, 258
turmeric 256
twin studies 53, 63, 64, 149, 155, 237, 251
tyrosine 114

Uber Eats: 'Cravings Report' 211, 231
United Nations (UN) 101, 151, 168, 171
University College London (UCL) 167
University of Amsterdam Medical Centre xix, 227
University of Chicago 154
University of Michigan 225–6
University of Puerto Rico 152
University of Washington 236–7; Department of Global Health 179
urbanization 40, 151–3, 157, 246

vaccination 14, 42, 43, 45, 58, 77, 118, 162, 182, 185–9, 190, 260, 272, 273, 276
vagina 9, 50, 75–8, 83, 84–5, 87–8, 97–8, 106, 174, 249, 259, 264; bacterial vaginosis 75–8, 259–60
vagus nerve 109, 112, 113, 119, 138, 219, 239
vancomycin xviii, xix, 177, 179, 183
variola virus 42
vasoactive hormone 83
Vedic civilization 176
Veillonella 117, 131, 132
Veillonella atypica 266

Veillonellaceae 265
Venter, Craig 11, 271
vertical transmission 91–2
villi xxii, 28, 65
viruses 10, 15–16, 44, 47, 48, 49, 57, 60, 312; Clustered Regularly Interspaced Short Palindromic Repeats (CRISPR) 16, 172; definition of 10; evolution of 23; immune system and 16; microbiome, shaping of 15; numbers of 15; phages *see* phages; sexually transmitted infection (STI) and 77–8 *see also individual virus name*
vitamins 34, 86, 93, 94, 100, 138, 206, 220, 222, 223, 245

Waksman, Selman 177, 181
WarGames 55–6, 57, 59, 61, 62, 68, 69
Warren, Robin 47
Watson, Jim 163
Weizmann Institute of Science 138
West Germany 54–5, 62, 64, 67, 68, 140, 152–3, 170
wheat 32, 159, 216, 220
white blood cells 57–8, 96, 128
Woese, Carl 8, 9–10, 17, 20, 273, 276
Wolbachia pipientis 80
World Health Organization (WHO) 183, 186, 228, 246, 262
World Organization for Animal Health (WOAH) 172
Wuhan University, China 85
Wulfstan, Archbishop 3

xanthan gum (XG) 225–6
Xanthomonas campestris 225–6
xenobiotics 94, 100, 254

Yersinia pestis 41
York Gospels 3

Zola, Signor 198–9
zombie ant 121–2
zoonotic transmission 40–41, 166
Zuboff, Shoshana 271
Zwicky, Fritz 6, 17, 276